Deerproofing Your Yard & Garden

Deerproofing Your Yard & Garden

RHONDA MASSINGHAM HART

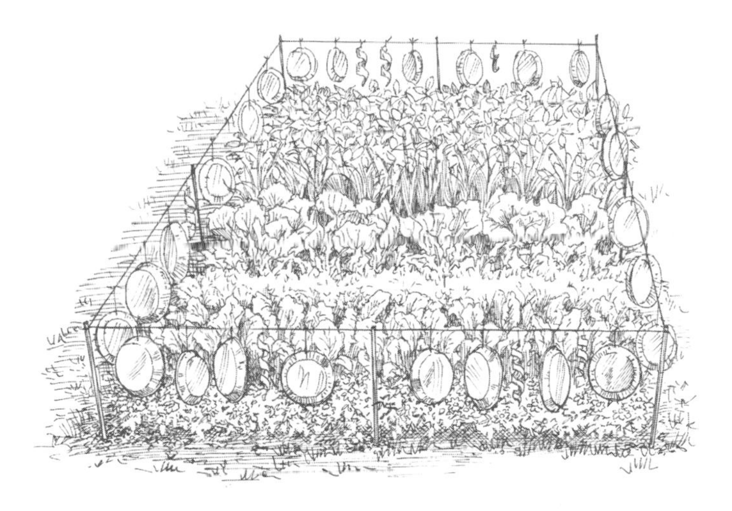

Storey Publishing

The mission of Storey Publishing is to serve our customers by publishing practical information that encourages personal independence in harmony with the environment.

Edited by Charles W.G. Smith, Gwen Steege, and Carleen Perkins
Cover design by Kent Lew
Cover photograph by © Janet Haas/Index Stock
Text design by Jennifer Jepson Smith
Text production by Jessica Armstrong
Illustrations by Elayne Sears
Indexed by Jan Williams

Printed in the United States by Versa Press
10 9 8 7 6

Library of Congress Cataloging-in-Publication Data

Hart, Rhonda Massingham, 1959–
Deerproofing your yard & garden / Rhonda Massingham Hart.
p. cm.
Includes bibliographical references and index.
ISBN-13: 978-1-58017-585-2 (alk. paper)
1. Deer—Control. 2. Wildlife pests—Control. I. Title: Deerproofing your yard and garden. II. Title.
SB994.D4H375 2005
635'.0496965—dc22 2004025042

CONTENTS

Acknowledgments

When Gwen Steege of Storey Publishing called to ask if I was interested in revising this book, my first thought was "what has changed?" You might be surprised. Deer populations have continued to grow over the past 10 years and the damage they inflict has increased exponentially. While we remain aware of the consequences of encroachment on deer habitat, a lot of people are desperate to protect their own habitat. So we agreed, an update could really help a lot of people.

Then she asked how I'd feel about sharing the rewrite with a thorough researcher and accomplished writer — Charles W. G. Smith. She felt he would bring a tremendous amount of insight and fresh perspective to the project, and was she ever right! What a wonderful blessing his input proved to be. Charly and I worked for months, chasing down promising new ideas, many of which turned out to be nothing more than flagrant exploitations of gardeners' fear and frustration, a few of which show genuine merit. We challenged every deterrent, so-called deer-resistant plant, and new product, until we narrowed down the list to those that have proven themselves in the most circumstances.

If you've ever participated in a big pot-luck dinner, you already know what it takes to put a book like this together. Everyone brought something valuable to the table. Gwen set us up for a feast, Charly and I brought the meat and potatoes, and dozens of others pitched in all manner of tasty items. Numerous gardeners, extension agents, researchers, and wildlife biologists from around the country contributed their specialties until we had a veritable banquet of ideas. Finally, Carleen Perkins, whose wit and good humor served as the perfect dessert after gorging ourselves on all that information, made it all digestible with her concise editing. My sincere thanks to all who made this spread possible.

Finally, no one deserves acknowledgement for enduring the process of this revision more than my daughter, Kailah Hart. She heard nothing but "deer talk" and saw nothing but the back of my head (as I was typing away) for months. Thank you, Munchkin, for putting up with me.

Foreword

by Jim Wilson, former host of *The Victory Garden* on PBS

For more than 25 years, my gardens have been ravaged by deer. I suppose my situation vis-a-vis deer is pretty typical. Our log-cabin home stands on 15 acres bordering a university town in central Missouri. From a hilltop near our home, we can see the campus; between it and us is a solid carpet of housing. Behind us, farmland stretches seemingly forever. Not surprisingly, we have resident deer. A doe and two fawns, now yearlings, are so accustomed to us that they hardly look up from vacuuming birdseed when we peer at them out our back windows.

We have about a quarter-acre in flowerbeds and a new 1,000-square-foot fenced food garden. To date, we have been able to protect the plant species most attractive to deer — hostas, daylilies, phlox, etc. — by regular spraying with Liquid Fence. Our decorative fence around the food garden is not high enough to keep out deer, so we are considering ways to correct that shortcoming.

Gardeners like me are all too aware of the damage that can be wrought on landscapes and food gardens by deer. Some also understand the profound, long-term damage to local ecosystems by hungry deer that are driven to foraging on forest-floor wildflowers. Years ago, Lori Otto, a legendary environmentalist from Milwaukee, Wisconsin, walked the woods near her home with me and pointed out the almost complete extermination of ephemeral wildflowers. Without these early-blooming flowers, early hatches of insects have no food, and early-arriving, neotropical, largely insectivorous birds have little or no sustenance to fuel their northerly migration. Without insectivorous birds, insect pests can multiply uncontrollably in our gardens, fields, and forests.

Society has exacerbated the deer problem by allowing, even encouraging, urban sprawl, which has extended homes and gardens into country that once sustained deer. Gardeners have unwittingly magnified the problem by feeding and irrigating lawns and gardens. Especially during dry weather, when grasslands and wild brush have withered, deer will be drawn to the succulent feasts laid out for them by homeowners.

Long ago, I should have studied the habits of these pesky critters, so I could have done a better job of outsmarting them. But I never came across a definitive reference book on the essential nature of deer. Oh, I had read many anecdotal recommendations on how to control them, but most of the advice contradicted my own experience. So, I adopted a "Sez who!" attitude . . . until now, that is.

When Rhonda Massingham Hart researched this book, she did the next best thing to pulling a deerskin over her clothing and mingling with herds. First, she consulted zoologists and game management specialists to understand the social structure of herds, feeding and mating habits, seasonal variations, and deer–human interactions. Then she interviewed horticulturists at public gardens who have to cope with deer day in and day out, all year long. Finally, she put it all together in an honest, thorough, highly readable book that is appropriate in every corner of the United States and southern Canada.

Other than very high fences, *Deerproofing Your Yard & Garden* doesn't offer a universal recommendation for protecting gardens from damage, but it does offer you a sampler of strategies that work most of the time under most conditions, and it gently but firmly pooh-poohs ineffective gadgets and panaceas. Thanks to Ms. Hart, I will be considering alternating various kinds of repellents on my flowerbeds and will be trying one of those proximity-sensitive sprinkler heads for my food garden. I plan to reach out, not up, with an extension on my food garden fence.

I have been quietly lobbying for a dog to replace our dear departed Sheltie, despite Ms. Hart's sensible observation that dogs tend to sleep when deer are most active. But bars of soap? No, not after seeing deer tooth marks in bars of Irish Spring hung at the end of crop rows at Callaway Garden when we were taping segments for *The Victory Garden* years ago.

Yes, Ms. Hart, you can teach an old dog new tricks, and this old dog thanks you. Arf!

ONE

Not Your Ordinary Problem

On a good day, this is an uncertain age. It seems that every morning dawns to new stresses, from shaky economics to alarming international politics. Life can become so overwhelming. But even in these hectic times, you can count on having at least one sanctuary, a world of your own design and composition, a retreat where turmoil stays outside and peace, solace, and even a measure of control are inside. That sanctuary is your garden.

Our gardens are unique in that they are at once our own creation and a covenant with the earth. Here we have the power to plant and nurture, the ability to foster new life and beauty and to find solidarity with the natural world. Here the outside world dissolves as we till and toil, work the soil, compost, weed, and fill our souls with the scent of earth and green.

That is, until the day when something is amiss. Plants are damaged and some are missing. The next morning, the damage is worse, and by the third, destruction is everywhere. Is nothing sacred? Not to deer.

Deer have only two things on their minds — survival and reproduction. They display neither benevolence nor malice, just an ongoing healthy appetite for much of what grows in your garden, including your tasty tulips and yummy yews. Deterring deer from your little patch of paradise is not your ordinary gardening problem. But that's okay, because the solutions in this book aren't ordinary either. So when your safe haven comes under attack, you'll have ample ways to defend it.

Deer can easily clear a four-foot fence.

The Scope of the Challenge

As you take up the gauntlet of protecting your garden or landscape, try to take some comfort in knowing that you are not alone. Boy, are you not alone!

A funny thing happens when people move into deer country — the deer don't go away. For all our technological prowess and progress, it often appears that they have us surrounded. Accurate national statistics are next to nil (deer are notorious for not returning their census questionnaires), but state-by-state reports are plentiful. It's a tad ironic that deer were all but extirpated from many regions only a century ago (they were virtually gone from Indiana, and only 400 were believed to exist in Missouri by 1925). The deer population has rebounded to an estimated 20 million, which now threatens our collective sanity. That's a lot of competition for our garden greenery.

Deer may not be everywhere, but just like a certain credit card, they seem to be everywhere you want to be. Historically, about the only gardens guaranteed to be deer free were in Australia and Antarctica. However, in the late 1800s many deer species — including fallow deer, the chital, hog deer, red deer, rusa, and sambra — were introduced to Australia, where they now thrive. That leaves only Antarctica, and the growing season there is very short.

Throughout the United States and Canada, deer now thrive in the wild, but they also adapt insidiously well to suburban and even some urban areas. Wildlife biologists estimate that on average, a healthy natural environment should support from 18 to 24 deer per square mile, yet some areas are inundated. Parts of New Jersey have deer counts of 124 per square mile, South Carolina 76, Texas 113, and areas in New York approach 200. Just my yard in northeastern Washington state hosts between 12 and 30 on any given evening.

But really, can something with such big brown eyes, delicate features, and soft downy fur really cause all that much trouble? You betcha!

Damage by the Numbers

A single adult deer consumes between 6 and 10 pounds of green stuff every day, or about half a ton of plants over the growing season. An adult moose, the largest member of the deer family, can down 33 to 44 pounds of forage a day, or 2.7 tons of plants in just one summer. That's a lot of weeds, broad leaves, buds, and blossoms. Multiply that by the number of deer in your area, and it becomes clear why deer so often stray your way in search of a buffet.

True, some regions have more deer pressure than others. People in the Great Lakes region (Michigan, Wisconsin, Illinois, Indiana, and Ohio), northeastern states (Maryland, Massachusetts, New Jersey, Pennsylvania, New York, Connecticut, and Vermont), and parts of the South (especially Alabama, Mississippi, Georgia, and the Carolinas) endure the worst from deer. But none of that makes the gardener in Oregon whose roses disappeared overnight feel any better. If deer are in your garden, then your garden has a problem with deer.

A look at state agricultural reports shows that deer damage to commercial crops is estimated in the millions of dollars. New

Tulips are a favorite deer snack.

Jersey farmers estimate that $10 million worth of deer damage to crops occurs statewide. Indiana's farm economy reports a $13 million loss, and Maryland figures its loss is closer to $14 million annually, for just corn, wheat, and soybeans. A report published in 2002 by the National Agricultural Statistics Service (NASS) of the U.S. Department of Agriculture (USDA) credits deer as the leading cause of all wildlife-related field-crop losses, accounting for 48 percent. And that doesn't even begin to account for the damage to private yards and gardens.

Deer (and their hungry relatives) also wreak havoc in other ways. Susan Nottingham, a rancher in Burns, Colorado, lost 300 tons of hay to a herd of elk that moved through her ranch one December. In areas of the country where white-tailed deer are prevalent, such an invasion not only would have depleted the feed supply, but also might have left behind spinal meningeal worms or brucellosis in the herd. Deer–car collisions are another menace. Such accidents kill more people each year in the United States than collisions with any other animal, racking up national automobile insurance claims of well over $1.2 billion each year.

Solving the Problem

Clearly, the problem with deer is bigger than your backyard or mine. But how do we solve a problem that has been growing all around us for 100 years? We look at the problem realistically. In some situations, we can solve the problem; in others, the problem can be successfully managed. Whatever the predicament, there is always something we can do.

Deer can be managed humanely and effectively, and can be trained to coexist in our midst. It may not be possible to eliminate *all* deer damage *all* of the time, but rest assured, you can definitely incorporate strategies to make it less common and less destructive. There is no one simple answer for everyone, as so much depends on your individual circumstances, the deer pressure in your area right now, the time of year, your yard, and the property surrounding yours. Environmental factors like drought, mast crops, winter temperatures, and snow cover also affect what gets eaten and when. The trick is to learn what works in your own situation, and why it works. That way, when something changes — be it the weather, deer behavior, or some unknown factor — you will be able to adapt as quickly as the deer do.

The good news is that there are effective strategies for controlling deer behavior and curtailing the damage they cause. Commercial farm economies, with millions of dollars at stake, have set the stage for research into controlling deer damage, and have come up with solutions that really work in those settings. One alfalfa farmer near where I live spent thousands to erect a deerproof fence all around his field. Imagine his elation when he realized it had paid for itself with the first harvest. Professional landscape designers, under the gun to perform for their clients, create garden plans that minimize deer damage. University-funded studies apply what is known about deer behavior to further fine-tune these strategies, and success generally begets new funding for continued research. And thousands of gardeners throughout the country, just like you and me, experiment with new ideas and old on a regular basis, out of sheer desperation to protect our pansies.

As you can see, you really aren't alone. What deer are doing to your garden they have done to someone else's. In most cases, the problem has a solution. This book draws from all of the above resources to cut the fluff and deliver the best, most reliable solutions that work in the broadest circumstances. Some of them will work for you.

In short, this book not only educates you about deer, but also empowers you to take back your garden, sanely and sensibly, so it will always be your sanctuary.

TWO

Getting to Know Deer

Deer are extraordinarily adaptable creatures, both as individuals and as a family. Scientists and gardeners alike recognize their ability to adjust to any given set of circumstances as second only to that of rats. Their versatility spans ingenious behavioral adaptations, such as learning to sneak through or around fences, gauging the length of a chained dog's reach, sending and reading signals simply by flashing their tails, and blissfully ignoring traffic while they munch on median divider roses. The physical adaptations of deer include a sophisticated winter survival system, complete with luxurious, hollow-haired fur coats and a metabolism that slows down, allowing them to survive for long periods without eating. Another survival adaptation is a four-chambered stomach that digests just about anything.

This is not good news for gardeners trying to deal with them. How, then, does one even begin to limit their comings and goings? The first step in controlling deer is getting to know them.

Into the Wild

People tend to have one-dimensional perceptions of deer. Some see an image of ethereal grace, velvety spirits that fade in and out of the wild wood to honor us with fleeting glimpses of their august beauty. Others find them intolerable nuisances, rats on hooves. Still others regard them simply as the spoils of the game, trophies to attest to one's hunting skills. And then there are those of us who admire them in their natural state, enjoy their presence, tolerate minimal damage, eat venison on occasion, and don't faint at the sight of buckskin.

Each of these points of view has its merits and shortcomings. Deer are not one-dimensional creatures. They are not Bambi. They are not simply garden pests. They are far from "velvety spirits." Deer are flesh-and-bone wild animals whose struggle to survive is a daily life-and-death drama. But they wouldn't have survived so well and so long if our misinformed perceptions of them were true — they're smarter than they look. Nature has endowed them with far more fortitude and ingenuity than we may want to give them credit for.

Deer are well-equipped to avoid predators, but gardeners can take advantage of deer weaknesses to protect their gardens.

For eons, nature has endowed deer with an eclectic assortment of wondrously effective traits, most of them adaptations to detect and avoid predators. From the tips of their ears to the points of their hooves, deer are engineered to escape attack from natural enemies that include cougars, bears, lynx, wolves, and even golden eagles. These same traits make it easy for them to find, raid, and escape your garden before you can finish your first cup of coffee in the morning. But thankfully, some of these traits have weaknesses that allow the crafty gardener an avenue of recourse.

SENSE OF SMELL. Their first line of defense is a big wet nose — eight times bigger than that of a human, three times larger than that of most dogs. Deer rely heavily on their ability to detect and evaluate scent. A deer's sense of smell is also crucial for finding and identifying food, following deer trails, and recognizing its young. That super-sensitivity to smell may also be a deer's Achilles' heel when it comes to odors employed to repel them, as we'll see later.

SENSE OF HEARING. Another major asset is superior hearing, which, like a deer's ability to smell, far exceeds that of humans. Their big ears flex and rotate constantly to detect every sound, near and far, as quiet as the crushing of a dry leaf underfoot. They serve as an early warning system should you attempt to sneak out into the garden and catch them in the act, but they can also be used to fool the rest of the deer into fleeing from perceived dangers.

SENSE OF SIGHT. Eyesight is also important to deer. They can spy a tasty garden from a half-mile away, day or night. Large, prominent eyes register the slightest movement, even in dim light, thanks in part to a tapetum that reflects and thereby doubles the amount of light available to the retina. The placement of the deer's eyes on its head allows for roughly a 310-degree field of vision: binocular (both eyes focusing) in the front, monocular in a wide arc on the sides and toward the back of the head. Although this

Range of vision

The placement of a deer's eyes allows it to scan roughly 310 degrees.

arrangement makes it easy for a deer to constantly scan its horizons for predators, it makes it difficult for it to focus on any one object. Movement instantly gets a deer's attention, but a still object may not be detected. In exchange for superior night vision, a deer's perception of color is reduced. The deer's eye is loaded with rod cells, which function in low light, but has very few cone cells, which bring images into sharp focus and register color.

SPEED. A deer's best defense against predators is to run and jump. It has been said that a deer doesn't have to outrun the predator, just the other deer. When a deer's internal alarm goes off, its long legs — powered by strong muscles built for quick bursts of speed — and its sturdy, cloven hooves — designed to ensure traction in a wide range of footings — launch the deer away from danger. Deer can sprint as fast as 35 miles per hour and jump nine feet high. (Deer farms in Texas report that in order to keep bucks from does during breeding season, the fences must be 12 feet high!) They can also clear a span 20 feet across in the blink of an eye. Their narrow bodies and thin legs are built for agility. An antlerless deer can dive between two strands of barbed wire on a dead run. This athleticism makes them all the harder to exclude from the garden. If only deer were more like sloths!

NATURAL CAMOUFLAGE. A deer's coat color provides additional protection. The color, which varies with the habitat and the season, helps deer blend into the background. Deer native to dry, cold, open areas tend to be lighter in color than those that inhabit warm, humid regions or areas of deep cover. In cold weather, the slick summer coat sheds and is replaced with a duller, grayer shade for winter, permitting deer to fade more easily into the dreary backdrop of the season. The hairs of the winter coat are hollow, making it fluffier and more insulating: all the better to survive the winter snuggled securely in your yard.

Deer 101: An Easy Guide to Garden Deer

Before you can prevent deer damage, you have to get to know the enemy. The first step is to identify which species of deer (or other relative) is living in your area and what its habits and food preferences are. Once you have a good understanding of how deer behave, you'll stand a much better chance of keeping them out of your garden.

White-Tailed Deer *(Odocoileus virginianus)*

Easily identified by those large white tails that give them their name, the white-tailed deer (also referred to as a whitetail) is the nemesis of countless eastern gardeners. Acclimation and hunger, especially in regions where they are most heavily populated, have led many to overcome their natural shyness. Whitetails show the greatest variation in size, ranging from the tiny Florida Key deer (*O. v. clavium*), which averages a diminutive 28 inches in height and weighs in at around 80 pounds, to monster northern-woodland whitetails (*O. v. borealis*) that top 300 pounds. This variability reflects differences in climate (smaller bodies dissipate

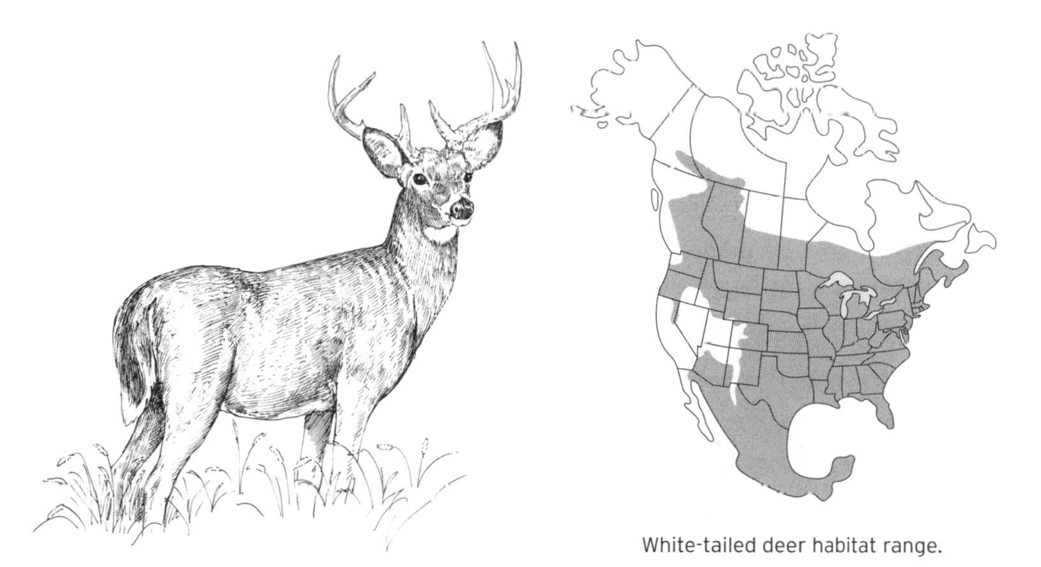

White-tailed deer habitat range.

heat more efficiently; larger bodies conserve it), a limited gene pool that fosters inbreeding (subspecies are limited in their growth potential), and nutrition (insufficient or inferior forage, which impedes growth).

Whitetails, though somewhat discriminating in their taste preferences, will devour plants that most other animals, including blacktails and mule deer, won't touch. Their preferred habitat is the classic woodland edge, but forays into suburban and urban areas have become all too common.

Population Density Chart

The number of deer per human inhabitant varies greatly from state to state, and often indicates areas where the natural environment is under pressure.

State	People**	Deer*	Number of people per deer
Alabama	4.5 million	1,500,000	3:1
Illinois	12.6 million	700,000	18:1
Iowa	2.9 million	325,000	9:1
Michigan	10 million	1,900,000	5:1
Mississippi	2.8 million	1,750,000	2:1
Montana	917,000	250,000	4:1
New York	19.2 million	1,000,000	19:1
Ohio	11.4 million	475,000	23:1
Texas	3.7 million	22,100,000	6:1
Wisconsin	1.6 million	5,400,000	3:1

* Deer populations compiled from various sources.
** Human populations from U.S. Census Bureau.

Mule Deer *(Odocoileus hemionus)*

Slightly larger-bodied than the average whitetail, mule deer also tend to weigh in a little heavier. The most common subspecies, the Rocky Mountain mule deer (*O. h. hemionus*), has a darker coat than the more typical medium brown. The desert mule deer (*O. h. crooki*) is another notable exception; its coat is a much paler shade of grayish tan, allowing it to blend into the surrounding desert.

Commonly called a muley, the mule deer gets its name from its huge ears. Big ears are a boon in areas where cover is scarce and predators may stalk from great distances. The ears also help to dissipate heat. Muleys have a bounding gait, called the stott. When stotting, the deer uses all four legs to bounce and then lands on all four feet at once. This allows the deer to change direction with each bound — a handy way to avoid predators, specially designed for their preferred habitat of rough, brushy hillsides. Mule deer can cover up to 25 feet on each stott and reach speeds of 25 miles per hour. By nature, muleys are less nervous than are whitetails. When nervous, they walk stiffly in contrast to their normal fluid movement, and are more apt to confront predators, including humans. Because they are less likely to run, they may seem relatively "tame."

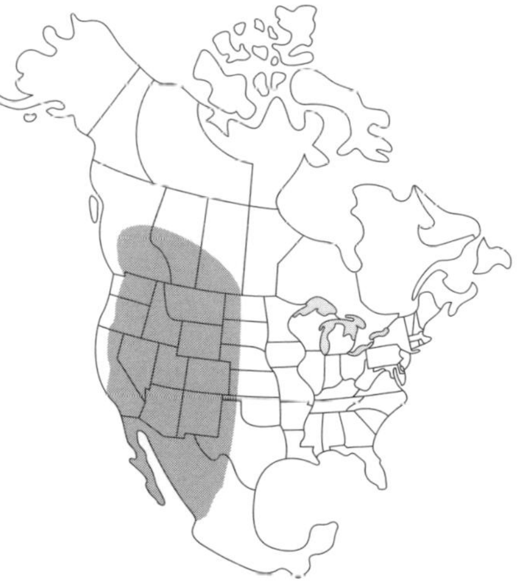

Mule deer habitat range.

Black-Tailed Deer *(Odocoileus hemionus)*

Native to the Pacific Coast are two subspecies of black-tailed deer, the Columbian (*O. h. columbianus*) and the Sitka (*O. h. sitkensis*). They prefer the cover of dense vegetation common in their native range. The Columbian is the larger of the two and the more likely to find its way into gardening territory, as the range of the Sitka blacktail is limited to the rain forests of the extreme north coast. This means that only gardeners from northern Washington south through much of Oregon have to contend with them, and, generally, only those in more rural areas. Wildlife biologists used to categorize blacktails as a subspecies of mule deer, but in 2002, using mitochondrial DNA, they established a surprising link among whitetails, blacktails, and mule deer. It was discovered that, rather than the blacktail descending from the mule deer, the mule deer species actually arose as the result of crosses between blacktail bucks and whitetail does at least 10,000 years ago.

Black-tailed deer habitat range.

Elk *(Cervus elaphus)*

Elk habitat is so restricted that most readers may wonder why it's necessary to mention them at all. These people don't live with elk and have never had to salvage a garden in their wake. Most vulnerable are yards, gardens, and orchards in Washington, Oregon, California, Idaho, parts of Montana, Utah, Wyoming, Colorado, and parts of Arizona and New Mexico. Many states now compensate farmers (and some individuals) for damages caused by wildlife. As of 2004, the state of Washington has $50,000 available to compensate farmers for damage to crops and fences caused by elk. The state of Colorado spends upwards of $200,000 to compensate for damages and another $370,000 for prevention of wildlife damage, including fencing to keep elk from crops. And these figures don't take into account damage to major corporations — namely, timber companies — or home gardeners.

Elk migrate in groups from small bands to enormous herds, and when they descend on your plot, they leave it looking as though a demolition crew just ran through. Fences are knocked over and entire gardens are trampled and devoured. In areas where elk are a problem, they can be a BIG problem.

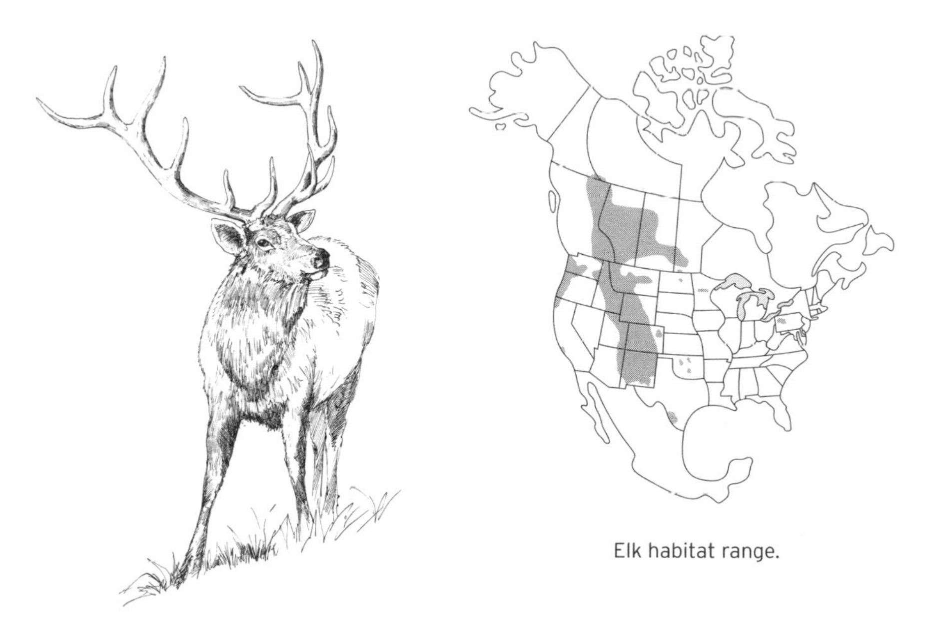

Elk habitat range.

Moose *(Alces americana)*

A mature moose can stand over seven feet tall at the shoulder and weigh from 1,300 to 1,500 pounds, with a few straining the scales at close to 1,800. Males sport racks of up to six feet across, weighing an average of 50 to 60 pounds. They feed primarily on browse, fir, aspen, and elder, and will eat their own weight in greenery each month. A moose in your backyard outclasses every other kind of deer problem you can imagine. Moose are the largest member of the deer family and the problems they create are just as big as they are. They are very big and very hungry. They are rude and often cranky. Reports of moose charging people are practically commonplace. Never try to chase away a moose; he might return the favor.

Moose are common in Canada, Alaska, and the far Northeast (including Vermont, New Hampshire, and Maine), as well as in limited areas of northeastern Washington, northern Idaho, western Montana, and south through Wyoming and Colorado. When snow covers the ground, they will gravitate to plowed or cleared areas, because it makes their job of excavating for forage so much easier. Other things that attract moose are their favorite

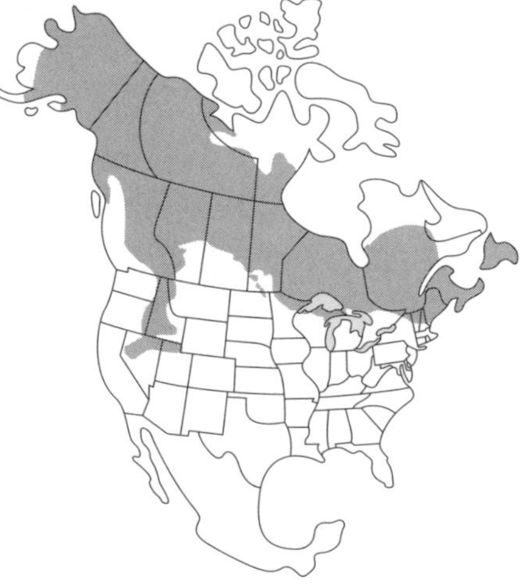

Moose habitat range.

Comparative Deer Family Characteristics

Deer	Average Size*	Color & Markings	Tail	Antlers
WHITETAILS				
Bucks	160 lbs./ 40 inches tall	Belly patches, white around the eyes	Brown on top, white underneath	Single beam, multiple tines
Does	120 lbs./ 38 inches tall	Tan, cinnamon, brown; white muzzle, throat		
MULE DEER				
Bucks	200 (rarely up to 400) lbs./ 42 inches tall	Dark to medium brown	Brown at top, white midsection, and black tip	Multiple forks
Does	140 lbs./ 40 inches tall			
BLACKTAILS				
Bucks	150 lbs./ 38 inches tall	Reddish brown to tan	Black with white underside	Multiple forks
Does	115 lbs./ 36 inches tall			
ELK				
Bulls	500–1,000 lbs.	Deep reddish brown, shaggy mane, tan or light-colored rump patch	Small, same color	Branching tines from a main beam; six points on average
Cows	400–700 lbs.			
MOOSE				
Bulls	1,500 lbs.	Light underbelly, dark brown hair with black tips	Primarily black with lighter points	Small; to 6 feet wide; palmate
Cows	1,300 lbs.			

* The size of deer varies with latitude and quality and quantity of food. The farther north and/or the better the food supply, the larger the deer.

foods — crab apple trees, roses, raspberries, and other related trees and shrubs, as well as natives like willow. Avoid planting such delicacies near your house if you live in moose territory, and never intentionally feed them; it encourages aggressive behavior.

Understanding Deer Behavior

Although it may well *seem* as if deer behavior is geared strictly toward destroying your yard or garden, what actually drives them is simply the will to survive. Some of the frustrating and inexplicable things they do — such as congregating in your yard in the winter, raiding your garden while ignoring your neighbor's, or snipping off the tops of all your bush beans rather than just eating a few whole plants — actually have reasonable explanations — if you're a deer. But buck up! Once you have a basic understanding of what motivates deer behavior, you will have a much better chance of modifying it to your advantage.

How Deer Think

Except during mating season, when the procreative drive takes over, deer live by the accompanying checklist. Everything a deer does, from the way it digests its food to the way it defines its territory, is designed to keep it from being devoured. Next in importance comes eating — no surprise to anyone needing this book! It also stands to reason that conserving energy and resting under cover are important parts of a deer's day.

Thus, while it may appear that deer spend more time thinking up ways to violate your violets than you do planting them, the reality is that they spend their days instinctively reacting to their environment. What passes for craftiness on their part really amounts to a good set of genes that have adapted well to the environment over many generations. The bad news here is that instinct is a difficult foe to foil. Once you understand the cues deer are responding to, however, you can manipulate them. Though it is a popular belief that a deer's instinct takes a backseat to its thinking ability, research by behavioral scientists comes to an opposite conclusion. Wild animals, including deer, are almost if not entirely driven by instinctive behaviors.

Additionally, deer have a social ranking that is important enough for them to expend considerable hard-earned calories maintaining. A dominant doe will physically challenge other does for the right of leadership by rearing up and thrashing them with her sharp hooves. Come mating season, a dominant buck — a mature buck with superior antlers and an attitude to match — will fight any buck in the woods for the supreme right to breed.

A Deer's "To-Do" List

A deer's "Things to Do Today" list is always the same, in this order (except during mating season):

- ✔ Don't get eaten
- ✔ Eat
- ✔ Rest
- ✔ Dominate other deer

Why should you care who gets to be King of the Forest? Because this social hierarchy also determines who is King of Your Garden. Once the dominant deer lay a path to your door, the others follow. Whitetails routinely follow established paths; muleys and blacktails test the way by scent. If others have gone before, then this must be the way to go! Once in the garden, it is the more dominant individuals that get first dibs on your daylilies and do the most damage. Less-dominant deer may be seen at the outskirts acting as sentries till their superiors have had their fill.

DEER TRUST THEIR NOSES. A deer's incredible sense of smell is its primary source of information. Glands on their bodies release various scents that deer "read" and interpret. Interdigital scent glands (between the toes) release scent wherever a deer walks. Other deer read these scent trails as the "all clear" for them to follow. A frightened deer gives off a distinct odor and other deer coming into the same area as the frightened deer, even hours later, show definite signs of distress.

DEER ARE CURIOUS, though always on the alert. They are especially curious about unfamiliar sounds. They will often cautiously approach a new sound rather than instantly run from it. Unless they feel threatened, which again boils down to how they instinctively read a thousand and one cues that most often remain undetectable to us, they tend to want to investigate new things that appear in their territory. Who knows — it might be good eatin'.

DEER RELY ON HABIT. It seems as though deer memorize every aspect of their territory and instantly perceive anything out of place. A strange sight or smell can immediately spook deer from an area. Conversely, once deer have accepted a sight, smell, or sound, they relegate it to the "normal" category and virtually ignore it. Deer will browse contentedly alongside highways (paying no mind to cars whizzing past), munch happily next to campgrounds amid radios blaring and children playing, even forage around military reserves during artillery practice, once they have become accustomed to these environments. Whatever association a deer learns to make with any given sight, smell, sound, or taste dictates whatever future response the deer will have to it.

DEER COMMUNICATE AND LISTEN. A lot is said through body language. The obvious example is the whitetail flag flying high and wagging from side to side in warning to other deer in the area. Deer that are nervous about possible danger assume a rigid posture and may stamp their front feet. Another warning signal is the snort, or sharp whistle — a shrill, sudden blast of air that sends scampering any deer within earshot. Deer also pay close attention to what other animals tell them. The squawk of a startled pheasant and the scolding calls of a squirrel mean just as

Six Keys to Deer Behavior

1. Paranoid thinking
2. Sensitive senses
3. Ironclad habits
4. Curiosity
5. Communication
6. Adaptability

much to the deer as to the animals that made them. And mating season has its own body language.

Aside from the big four on their to-do list, deer have few other needs. They generally won't stray too far from natural water sources — deer need from two to four quarts of water per day. But water is water, whether from a brook or from stock tanks, birdbaths, garden ponds, or fountains; it doesn't matter. Deer also favor salt or mineral deposits, including that salt block your neighbors put out for their kids' pony. Shelter is another concern, and in severe wind, hail, snow, or heat, deer will search out cover that may not meet with your approval. During the winter of 1988, we found four deer huddled beneath a pickup canopy that had been placed on blocks.

Extending the Invitation

Now that you have a clearer understanding of deer behavior, more needs to be said about how that behavior brings deer to your yard and garden. For that, we must consider dietary requirements and territorial imperatives in greater detail.

Eating Habits and Digestion

Deer are actually picky eaters — when they can afford to be. Though they can eat more than 500 kinds of plants, deer will hold out for their favorites when they can find them.

The deer's diet affects everything else in its life. An exceptional diet including a plentiful variety of lush, nutritious foliage, buds, and other forage develops large-bodied, robust deer; prompts does to give birth to twins or sometimes even triplets; and produces on bucks some of the most magnificent antlers imaginable. An adequate diet is one that leaves them with full stomachs, but full of food of marginal quality like wilted weeds or low-quality forage because of drought or poor soil. This will keep deer alive, enable does to conceive and bear a fawn or two, and allow bucks to grow a moderate set of antlers.

Malnourishment results in reduced fawning rates, underdeveloped antlers, susceptibility to disease, and, in extreme cases, slow starvation. It also forces deer to adapt their behavior and take serious risks in order to find food. The most common reason deer head for a new garden is hunger. If they are well nourished elsewhere, they generally don't seek out gardens; but, of course, there are exceptions. Sometimes a deer is just curious to see what you've got there, and then discovers a meal! Once deer learn where the food is — perhaps your hybrid tea-rose garden — they make a habit of returning.

Deer feed and digest their food in keeping with their prime directive, "Thou shalt not get eaten." They feed mostly during low-light hours, at dawn and dusk. Their eyes are well adapted to changing light levels; the eyes of most predators, including man, take as much as an hour to adjust. They eat "on the run" — a nibble here, a bite or two there, and then they move on. Sometimes they go just a few steps, sometimes a greater distance, but they don't feed in one spot for long. Deer also make a habit of feeding near some form of cover.

What about all the deer we've seen in the fields at dusk, nipping away at the tops of grass for an hour at a time? They are near cover, in that they know precisely where the shelter is located, whether it is the edge of the woods, a clump of trees, or a shallow swale. Usually, they will eat facing the prevailing wind, in hopes of catching any threatening scent in time to flee and reach cover.

Deer belong to a suborder of animals called ruminants, along with cattle, sheep, goats, and camels. They have a four-chambered stomach, which allows them to fill up the first chamber in about an hour or two and satisfy their immediate hunger. They can then regurgitate and chew their cud in peace and quiet. This greatly reduces the time spent in the open, potentially exposed to predators.

Territorial Limits

Within the vast range of area each species occupies, individual deer have their own personal range, or territory. This varies by habitat, species, season, and, to some degree, gender. The average territory of a whitetail is about one square mile, or around 700 acres. Territories may overlap, until the "rut" (breeding season), when bucks get serious about defending their area from interlopers. When not in rut, a buck may share or overlap his territory with other bucks; does tend to live in or near extended families of their own female fawns. Whitetail bucks and does live separately most of the year.

Deer territorial habits vary according to species. Whitetails are known for maintaining the same established territories for generations, even when the natural environment can no longer support their numbers and the sensible thing to do would be to move on. Blacktails tend to live in mixed groups in home territories that vary in size. Mule-deer bucks tend to have larger territories than does, with the size of the home range depending on the quality of the habitat. In some areas, it may be as small as two square miles, but in others, such as the Texas panhandle, one buck may range over an area as large as 25 square miles. Elk herds roam vast territories, which vary with the time of year. Whitetails tend to "yard up" (congregate) in winter; seasonal weather patterns routinely force mule deer and blacktails, as well as elk, to migrate from the high mountains to lower feeding grounds and back again.

Rogue Bucks

Romanticized though it has been, the story of the rogue buck vigorously defending his territory against all comers is fiction. Bucks do not maintain exclusive ranges. Their territories frequently overlap, and except during rut, there is little reason for conflict.

A deer's territory is bounded more by habit than anything else, for even in the absence of any physical barrier, little will tempt a deer from its home range. After all, if doing something or going somewhere the first time didn't get it eaten, then it's likely to be safe a second or third time. Familiar is as close to safe as it gets for a deer. Not surprisingly, keeping to the same territory is the main reason deer tend to raid the same gardens repeatedly — these gardens are in "their" territory.

When living in close proximity to human-populated areas, deer lose their fear of people.

The amount of land that constitutes a deer's home range depends primarily on the forage and cover available and on the number of deer in the vicinity. For instance, mule deer in the scrublands of Arizona and New Mexico must travel farther to find sufficient food than do whitetails in the rich farmlands of Illinois. In good forage, deer average a one- to two-square-mile territory, usually in a roughly elliptical pattern. If food is scarce or population density is high, they will roam much farther.

Natural and man-made boundaries can also determine where deer range. Distinct subspecies, such as the Hilton Head Island whitetail (*O. v. hiltonensis*) and the Blackbeard Island whitetail (*O. v. nigribarbis*), have developed on islands where the dividing waterways provided a natural barrier. In some places, major highways or urban developments form boundaries that limit deer territories.

Getting to Know You

There is a big difference between the deer of wild country and those living in the suburbs. Deer in the wild will flee at the mere whiff of human scent on a breeze. Deer in more urban areas will pull laundry off the line, steal food from picnic tables, and make themselves at home in yards, patios, even carports. Having become accustomed to the sight, sounds, movements, and smells we make, these suburbanites have lost their fear of humankind. To them we are "normal."

Deer that don't fear humans prove more of a nuisance than those we can scare away. They are certainly bolder when it comes to inviting themselves into our yards, gardens, and homes. I've seen mule deer come right up to a sliding glass door and stretch their heads inside to beg for bread. But they are still wild animals and can do considerable damage to themselves and their immediate surroundings if they panic and try to escape.

Perhaps the most ominous aspect of deer losing their natural fear of humans is that they may become aggressive. A mule-deer buck typically will hesitate in order to gauge an opponent and, if he decides he can "take him," will charge rather than retreat. Worse yet is the buck or bull in rut; no unarmed person should ever challenge a buck in rut.

Changing Times and Seasons

Changes in the season force definite changes in the daily routine of deer life. Spring brings new fawns, and sends last year's babies out into the world alone. During summer, does continue to nurse new fawns, bucks grow velvety antlers (the fastest-growing living tissue on earth) and keep to themselves, fawns grow, and young deer form groups that travel and feed together. Fall brings on the rut, during which all else is forsaken save the need to mate, and deer are much more active and unpredictable. Finally, winter brings on a slower lifestyle for whitetails and often a semi-migration for mules and blacktails as they seek out food sources. For the gardener, these behavioral adjustments mean a change of challenges, as all of these things affect deer's nutritional needs, and often they come looking to us to satisfy them.

Changes of Diet

Because seasonal changes bring with them distinct changes in the physiological state of deer, it should come as no surprise that these

changes also have a direct impact on deer's nutritional requirements. Though far from an exact science, deer nutrition has been studied at the university level and some interesting things have been discovered — some of which may demystify why the pansies in your yard are more attractive to deer than the obviously abundant native forage.

White-tailed deer have been studied the most, so exact figures may vary somewhat for other species, but we are at least in the ballpark with the basics. Most important, we have learned that nutritional requirements vary with gender, age, stage of growth, seasonal extremes, and physiological demands of pregnancy, lactation, and antler development. When it comes to deer pillaging our produce, this information helps us to understand what we are up against, when, and why.

During the spring and summer, deer face their highest nutritional demands of the year. Much of their life cycle depends on protein consumption. Young deer, including fawns and yearlings, require roughly 13 to 20 percent crude protein for growth. Lactating does have the greatest need for protein, especially those that are nursing twins. (Doe milk has been tested at 36.4 percent protein, which places a great demand on a doe's body, requiring a diet of at least 18 percent protein.) Surprisingly, bucks need roughly 13 to 16 percent crude protein in their diets during this period as well, because of the high amount of protein required to grow antlers. (Antlers comprise primarily protein-rich collagen and, once hardened, are about 45 percent protein.) All this explains, at least partially, why deer naturally seek out plants in nitrogen-rich or fertilized soils at a time when plant growth is at its peak. Nitrogen-fixing plants — those that convert nitrogen in the atmosphere through a symbiotic association with soil-dwelling rhizobia bacteria — are also highly sought after. This is what makes early peas and alfalfa, winter wheat and clover crops such prime targets early in the growing season.

Another way of looking at a deer's nutritional needs is to consider the total energy requirements. Overall caloric intake is measured in kilocalories and is called digestible energy (DE). DE is highest in the spring and summer. Does require about 33 kilocalories of DE per pound of body weight per day when nursing, compared to about 25 for body maintenance, or about 3,200 kilocalories per day for a 120-pound doe. This translates to a need for plants that contain 55 to 60 percent dry, digestible matter, much more prevalent in cultivated crops than in native forage. Rapidly growing fawns require more than twice the calories of their moms, up to 70 kilocalories per pound of body weight, most of which is supplied by their mothers' milk. During summer antler growth, it is estimated that bucks require 40 kilocalories per pound to maintain body weight and produce antlers.

It may be a surprise to realize that deer don't have rigid requirements for carbohydrates and fats; a deer's need for these dietary components is limited to the number of calories they contribute. As with other dietary components, these nutrients are most needed during spring and summer.

Mineral requirements are another somewhat gray area in deer nutrition, with different studies yielding conflicting results. We know that mature antlers are roughly 22 percent calcium and

Super Food for Deer

Acorns, prized by whitetails and blacktails, litter the base of oak trees and can be detected by smell even beneath a layer of snow. The acorn crop can have a direct affect on deer health and population. In years when acorn production is limited, does come into estrus later in the season, bucks come into rut later, and breeding activity is severely curtailed. Resulting fawns may be smaller and less hardy, and may mature more slowly, meaning a later reproductive cycle for them as well. One bad year can have negative effects for generations to come.

11 percent phosphorus. Yet various studies have shown different minimum dietary requirements of these elements — 0.09 percent to 0.64 percent for calcium and from 0.2 percent to 0.56 percent for phosphorus — to sustain antler growth. One reason for the wide range of findings in these studies is that bucks store minerals in their bones and undergo a sort of osteoporosis, which robs minerals from their bones to supply their antlers. The result is that their immediate diet is not their only source of minerals. After the antlers harden, the bucks replace through the diet minerals taken from the bone. This may explain why they congregate around livestock salt and mineral blocks in the fall.

As summer advances and growth, nursing, and antler development peak, so too do deer's nutritional needs. The available diet also changes, as plants begin to mature and flower or fruit, and broadleaf forbs develop. Although forbs and grasses remain the primary source of nutrition in the wild, legumes (peas and beans, for example) and maturing corn become more enticing in the garden. The sweetness and flavor of strawberries and peaches make them as attractive to deer as they are to people. And during summer, plants with high moisture content become a prime commodity. Well-watered crops or garden plants of almost any kind become sought after, as natural water sources begin to dry up.

Late summer through early fall is when deer really get serious about eating. All deer consume more food at this time of year, as they try to put on fat while being weaned (fawns), lactating (does), or growing antlers (bucks). During this time, available forage continues to change as well. Native plants toughen and become less palatable, yet well-tended garden plants continue to be tender and full of moisture. Grains, berries, and fallen leaves are plentiful.

When food becomes scarce, deer begin to branch out into evergreen boughs and practically anything else within reach. Cultivated crops, such as alfalfa, apples, soybeans, and winter wheat, along with the diversity offered by home gardens, constitute welcome additions to the diet, as deer try to put on enough weight to see them through the hardship of the coming winter.

As autumn wears on, nutritional demands change dramatically. Fawn growth slows down, as diminishing light triggers hormonal changes. Does nurse less or wean growing fawns that have long nibbled alongside them. And bucks may stop eating altogether as the rut takes over their brains and bodies.

Winter is the most difficult season for deer, particularly in the North. Food is scarce and the quality is low; only about 10 percent of what deer can find to eat is digestible. Though they browse on dead leaves, twigs, bark, and evergreen boughs when possible, they can actually starve to death with a full stomach. Blacktails and muleys may stray from home territories in search of food, but whitetails often form groups that "yard up," spending as much time as possible off their feet. (Standing burns nine percent more energy than lying down.) This time of year, their metabolism slows down and they tend to be very sedentary, unless forced to expend energy (being chased by dogs or harassed by gardeners, for instance).

Though the nutritional needs of deer change with the seasons, cultivated crops are always welcome additions to the diet.

Deer become all the more aggressive in their feeding habits, however, when there is less available. Even though deer have been known to survive for weeks without food, starvation eliminates fear and forces them to try things they wouldn't otherwise consider. This is especially evident in northern climates where winter can last half the year. Food items deer ignore during the lushness of spring, such as bark, twigs, conifer boughs, stacked cornstalks, and compost piles, may become preferred foods during the poverty of winter. Luckily, most plants can sustain the damage better if nibbled during this dormant period.

Spring Variables

- Florida whitetail bucks begin antler growth about a month ahead of northern bucks. The longer daylight hours trigger an increase in testosterone, which spurs antler growth.
- Desert mule deer grow antlers about two months later than other deer: Geared toward desert cycles, antler growth depends more on moisture than on day length.

Spring Ahead

As in all of nature, spring is a time of renewal for deer. Fawns are born and nursed. Antlers on bucks begin to grow. Shaggy winter coats molt, to be replaced by sleek spring finery. Surviving deer often emerge from winter in a state of near starvation and are anxious to regain lost weight. This bodybuilding requires lots of high-protein fuel, and deer will feed voraciously.

Elk, mule deer, and blacktails that had migrated to lower ground the previous winter now head back toward higher elevations. "Yards" of whitetails break up and the deer resume their normal territories. If your garden is in the path of these seasonal movements, it may become a prime stopover as tempting, tender new plants emerge from the soil.

Deer are reclusive in springtime. Pregnant does become restless and seek out a nursery territory, in as secluded an area as possible, with as much natural cover as possible, which they aggressively protect until their new fawns are about two months old. Except in the extreme Southwest and Southeast, most adult does give birth in May or June, typically to twins. But first, the

Orphans?

New fawns have no odor of their own, and does, to prevent predators from finding their young via their own scent, routinely leave new fawns alone during the day, bedded down in a safe place. There they wait until Mama returns for the eight o'clock feeding. They are not abandoned, as people sometimes believe, and should be left alone. Wild does may, however, abandon fawns that have human scent on them.

does chase away last year's fawns. These confused young deer are likely to turn up anywhere, including your yard and the highway. Keep an eye out for misplaced yearling deer in late spring.

Bucks stay in a limited area and are active for shorter periods, usually after dark. Their growing antlers are soft and loaded with blood vessels and sensitive nerve endings. The antlers grow quickly, about half an inch per day. Bucks know how easy it is to damage these immature status symbols and are reluctant to risk hurting themselves, so they take it slow and hide out.

The Good Old Summertime

Summertime and the living is easy — comparatively. Food is plentiful, at least in areas where deer don't overwhelm the habitat, and though not as rich as springtime fare, it is still nourishing. Fawns grow, learn to eat greens, and play. By the first of August, antlers reach their full growth and begin to harden, decreasing the nutritional demand for protein, calcium, phosphorus, and possibly other minerals in the buck's body.

By late August to early September, shortening day length triggers bucks to begin the ritual of assaulting small trees, both to remove the now dead, itchy velvet that once nourished the growing antlers (though this process takes only about 24 hours) and to advertise their presence to does with their distinct odor. These tree rubs are saturated with scent from glands in a buck's head; the more dominant the buck, the stronger the scent. Researchers believe that the hormones left in these rubs by high-ranking bucks can repress sexual behavior in less-dominant males, thus preserving the integrity of the population and the social order. Bucks in the mood will rub up on just about anything, but prefer saplings or shrubs from one to four inches in diameter, with smooth bark and no low limbs. Only older bucks attack larger-diameter trees, and will often rub the same tree repeatedly. In the Northeast and Great Lakes areas, quaking aspen is a favorite tar-

By late summer, bucks have begun rubbing off the velvet from their antlers. This signals the start of the "rut," which can last through December.

get, and in the Southeast, aromatic species, such as cedar, are preferred. Trees that have been skinned of their bark are a sign that deer are near.

Summer is not without its vexations for deer, however. The dense, short, solid hairs of the reddish brown summer coat protect the skin from sunlight and insect bites, but bugs nevertheless plague deer in hot months. If water is available, deer will plunge in to avoid them. To escape summer heat, deer become more active at night and head for cover during the daylight hours.

Autumn Daze

Because of the timing, we could say that deer "fall" in love. Or at least in lust. During the late autumn — mid-November for most of the United States, later in warmer areas (but year-round in tropical climes) — does come into estrus, beginning when they are about a year and a half old. The period in which they are receptive to breeding lasts only about 24 hours during this cycle. Compound

In a Rut

How can you recognize a potentially dangerous buck? In most regions, the first indicator is the time of year, but there are plenty of visual signals as well. Serious contenders for dominance sport serious antlers. Consider any buck with more than a set of spikes as a contender. Add a dark face to the criteria to watch out for. Glands on the forehead produce an oily substance that stains the faces of dominant bucks when they rub and battle against tree trunks.

As a buck goes into rut, his neck swells, making him look like a woodland linebacker. Bucks will often flare their neck hairs to make their necks look even bigger. When a buck is really looking for trouble, his hair stands on end all over his body, so he looks bigger and his coat looks darker. He will tuck in his chin and give his opponent a hard, wide-eyed look, ears flat back against his neck, antlers tipped toward his rival. In his anger, he licks his nose and flicks his tongue constantly. An angry deer's stance is rigid and he walks stiffly. He looks mad. Note that both elk and moose also exhibit similar rutting behavior.

Because a buck in rut may choose even otherwise unappetizing trees and shrubs as sparring partners, you should take additional precautions to protect trees and shrubs from damage. Young trees with narrow-diameter, smooth-barked trunks — such as cherry and willow — are particularly vulnerable.

A potentially aggressive buck during rut can be distinguished by his thick neck, flared hair, darker coat, and aggressive posture.

this with the fact that most does come into heat within a few days of each other, and it's easy to see why bucks go a little nuts.

Only dominant bucks have the right to breed, something established over the summer and early fall through jousting matches and displays of machismo. They drive themselves mercilessly in their search for just the right doe on just the right day, leaving dozens of scrapes and scent markers throughout their now greatly expanded territory; lusting bucks may patrol five times their usual range. They make the rounds of these markers constantly. A breeding buck loses from 20 to 30 percent of his body weight chasing does and keeping away subordinate bucks — this just prior to the onset of winter, a time when he can least afford the loss.

Testosterone levels soar and bucks in rut, wild and wily, can become dangerously aggressive. They are, in fact, more dangerous just before they begin to breed than they are in the thick of it. Though rare, bucks have been known to kill other bucks in dominance battles. They have also gored pets, livestock, and people to death and should never be approached indiscriminately during rut, which lasts seven to eight weeks in the North and longer in the South.

Enduring Winter

Deer experience a slowing of their metabolic rate as temperatures drop. They become sedentary, a life-saving state under the conditions. Virtually all growth stops, as maintenance becomes the goal.

Antlers — solid bone that could slash an opponent to pieces one day — suddenly demineralize and drop off, as the buck's body reabsorbs much-needed calcium.

Surviving Winter

As winter nears, deer scrounge for anything that looks fattening. Winter food holds little nutrition, and a thick layer of fat is a deer's best insurance that it will live to see another spring. Its next most-pressing need is to stave off the cold. A change of coats from the reddish brown of summer to the duller grayish brown of winter allows for comfort as well as camouflage. The long, hollow hairs of the winter coat hold in body heat close to the skin and provide excellent insulation. The darker color even aids in absorbing what little heat the winter sun provides.

The approach of winter pushes deer to seek refuge. Elk, mule deer, and blacktails move down from high-mountain feeding grounds in a mass exodus. Whitetails yard up, perhaps for the protection many eyes, ears, and noses can afford. Many feet also pack down snow, making it easier for the deer to move around.

Now weather becomes the biggest enemy. Windchill can make the cold unbearable, and sometimes unsurvivable. Deer seek out south-facing slopes to feed or rest, turning themselves into little solar collectors. During snowstorms, deer bed down and don't stir until the storm passes, even if that means waiting it out for days. Moose need deep, soft snow as insulation — when they bed down, they envelop their big bodies in the snow.

Snow multiplies the dangers of being a deer. Food becomes even scarcer, with twigs, bark, and dead plants buried beneath the snow. Evergreen boughs blown down by windstorms make welcome treats. Although deer will paw through several inches of snow for food, there is a limit. About eight inches usually discourages them from even trying, but they can sniff out some foods — apples, acorns, corn — under a foot of snow. Worse yet, snow makes it more difficult to move around; the deer's sharp hooves were designed for solid ground, not snowshoeing. Pushing through snow also burns more calories at a time when a deer has none to spare, and its markedly slower movements give predators an advantage.

As winter wears on, starvation begins to take its toll. Twigs and bark are stripped from ground level to as high as the deer can reach when standing on its back legs. Deer prefer smaller twigs (those no bigger than a wooden matchstick), and if larger twigs are being consumed, take it as a sign that the deer in your area are starving. If water isn't available, they will eat snow, which burns extra calories, furthering the progress of starvation. Deer become weaker and less active. If the winter is severe enough, an unborn fawn may be aborted or reabsorbed by the mother's body. Starvation gives diseases and parasites an open door, which can compromise the digestive system to the point that nourishment cannot be processed.

A starving deer stands with its back arched against the cold. If it can walk, it may stagger and fall. Its eyes are dull, its coat rough, its bones protrude through the dense hair. Where there is one such suffering animal, there are more. No longer the magnificent embodiment of grace and wild beauty, starving deer can only wait — for spring, if it comes in time, for death if it doesn't. It is not unusual for as much as half of a given deer population to die in winter.

THREE

What are the Damages?

As we have seen, because people and deer are sharing more and more of each other's habitat, deer damage has become a serious issue. The destruction they cause in our backyards is sometimes the most apparent. Deer can destroy a vegetable plot, strip berry bushes, and trample flowers in a single visit, leaving gardeners feeling violated, helpless, and frustrated. Elk and moose can also knock down fences, causing financial loss and setting livestock free.

But perhaps the most serious deer-related damage occurs outside the garden gate. Deer–car collisions do more than dent or ding; human fatalities are often a result. Interaction with deer can also cause the spread of disease to humans and other animals — most notably as carriers of the deer ticks that spread Lyme disease. Each year in this country, millions of gardens, thousands of vehicles, and untold numbers of pets, livestock, and people are affected by deer being in the wrong place at the wrong time.

Garden Damage

Right now, out there somewhere, a deer is gobbling down someone's prize posies or hard-earned produce. The estimated annual loss of crops, garden, and landscaping runs into hundreds of millions of dollars nationwide. Deer damage a wide variety of vegetables, fruit trees, nursery stock, and ornamental landscape plantings. The damage is not only immediate but also long term, reducing yields in crops and fruit trees and permanently destroying ornamentals and nursery stock.

The true cost of such damage goes beyond the financial or aesthetic setbacks. Most people garden, at least in part, because of a need to connect with the earth itself, to be part of the natural process of planting, nurturing, and harvesting. We garden to create our own private patch of Eden, to renew ourselves and escape the demands of daily life. There is a true, deep, emotional investment. When deer violate these living sanctuaries, they damage more than plants. They hurt us.

Investigating the Scene

Of course, deer aren't the only animals that damage gardens. There are many more — woodchucks, wild turkeys, rabbits, domestic animals, and then some. Start by taking a mental inventory of your surroundings. Just because deer live in your area, don't automatically assume that they are the cause of any and all garden damage. Knowing more about deer, as well as recognizing the clues they leave behind, can help you quickly determine if the damage in your yard is caused by deer or by something else.

Because deer tend to eat on the move, minimal nibbling need not be all that damaging to plants. Who will notice a few missing leaves? At other times, the effects of a deer raid are instantly evident. Flowers lie crushed and trampled, potted plants are overturned, and entire plants are gone or ripped up by the roots.

Whenever deer invite themselves for dinner, though, certain telltale clues remain. The first of these is what time of day the damage occurs. Deer are most active in the hours just after dusk and just before dawn (and black-tailed deer are even more nocturnal than other deer species). Damage done in broad daylight might lead you to suspect your neighbor's goat, or other wildlife. You can also evaluate the physical evidence — in the form of footprints, droppings, nesting, and even the type of damage left behind — but you may have to look carefully for it.

> We garden to create our own private patch of Eden. When deer violate these living sanctuaries, they damage more than plants. They hurt us.

Tracks and Signs

Left in soft soil, mud, or snow, tracks are usually excellent identifiers. Be aware, though, that other animals leave tracks that can be easily mistaken for those of deer. Elk and cattle tracks are similar, though elk tracks are more pointed at the tips than those of cattle. Llamas, too, leave tracks much like those of elk. Goat tracks can also look a lot like deer tracks. In relying on tracks to tell the tale, it helps to know the livestock in your area.

Deer tracks are usually about three inches in length. The track depends on the condition of the ground, the deer's speed, and the deer that left it. Determining when the track was made can be a little tricky, but if you make it a practice to lightly rake over paths or areas where you suspect deer damage and check the area every day, you'll know right away when fresh tracks appear. Generally, the sharper the edges of the tracks, the fresher they are. But many kinds of weather and conditions, from mud, rain, snow, and wind to freezing and thawing temperatures, affect tracks, making them look bigger, older, or fresher than they really are. Running deer leave tracks much farther apart than those of walking or browsing deer.

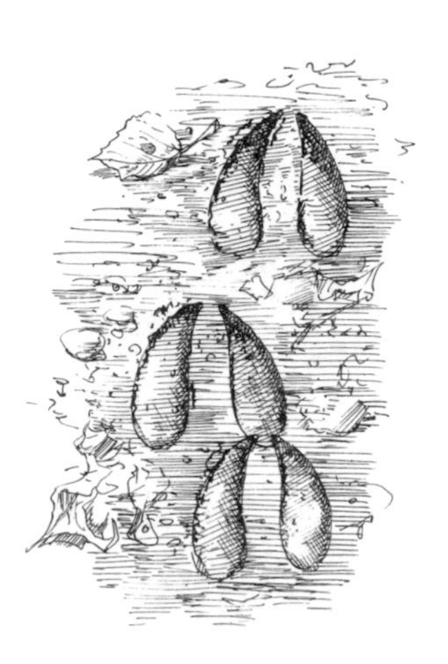

Buck tracks are slightly wider than those left by does, and if the soil is soft (or covered by less than an inch of snow), their dewclaws will often leave characteristic drag marks. Because bucks are heavier and more active than does, they tend to wear down their hooves faster, sometimes resulting in tracks that are more rounded at the tips. Bucks are also wider in the shoulders than in the pelvis, so the tracks from their front hooves will be slightly wider than those from their hind hooves.

In comparison, a doe's pelvis is wider than her shoulders, which results in slightly wider hind tracks than fore tracks. Also, during the summer, does often travel with their fawns, so doe tracks may be accompanied by one or two smaller sets.

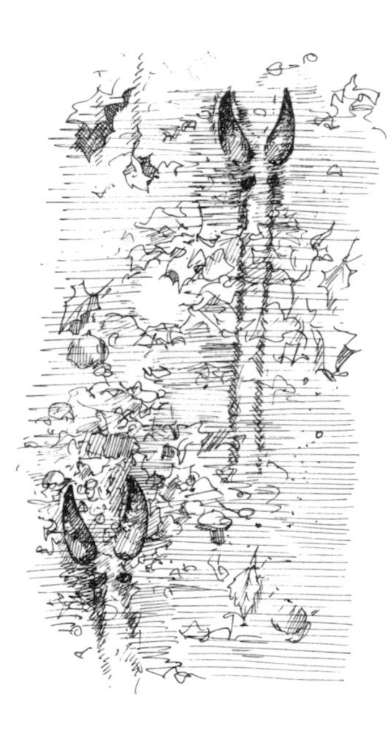

Deer tracks are about three inches long (above). In soft soil or thin snow cover, dewclaws of bucks may leave characteristic drag marks (below).

SCAT. Another calling card left by deer and other garden raiders is scat, or droppings. In late summer, fall, winter, and early spring, when a deer's diet consists mostly of browse (tender shoots, twigs, and leaves), the droppings are small, elongated pellets. When deer feed on lush greenery or fruit, the pellets clump together in a mass. Warm and shiny droppings are fresh, meaning deer are still very near. Cool and shiny droppings are no more than a few hours old, and dull droppings could be from days ago. Bucks tend to take potty stops; does tend to be constantly on the move. Small piles of scat, therefore, point to a

Raiders of the Dark

Many years ago a resident of Woodinville, Washington, felt certain that elk were regularly ransacking her garden. Every so often, she would find beanpoles shoved over and trampled, entire rows of produce munched to the ground, and, most incriminating, large cloven-hoofed tracks that had churned up her carefully tended soil. The raids always happened at night, so she had not actually seen the offending creatures.

Local officials advised her that she might be able to file a claim with the state for compensation, but she would have to prove that elk were damaging her property. Determined to catch them in the act, she devised a plan. She ran twine around the perimeter of the garden, with two or three tin cans tied together at irregular intervals. Surely when the elk trespassed next time, they would set off her alarm and she would have her proof.

Sure enough, three days after she set the trap, clanging cans alerted her that intruders had entered her garden. She rushed out to confront the culprits, only to find her neighbor's cow contentedly enjoying a midnight snack.

buck as the likely creator, whereas scattered droppings are probably the work of a meandering doe.

Unfortunately, even here impostors can confound things. Rabbit droppings look much like those of deer, but are always round and usually smaller. Goats, sheep, and llamas leave elongated pellets, similar to those of deer on browse. However, some of the usual suspects can be ruled out when pellet scat is your best clue. Pellets found at the base of trees with missing sections of bark and/or gnawed limbs are most likely from porcupines. Porcupines feed on the cambium — the layer between the bark and wood — of trees and shrubs, leaving damage that at first glance may be mistaken for a buck rub. Raccoons, skunks, opossums, woodchucks, marmots, and bears leave behind droppings similar to a dog's, which vary with the size of the animal and with the quality and quantity of their diet.

DEER BEDS. The presence of deer beds — packed-down, swirled sections of grass about three feet around — also indicates that deer feel at home in or near your yard. Deer bed down for the day in brush or tall grass, camouflaged from potential predators. If an area seems safe to them, they will make a habit of bedding down in the general vicinity. Having found a particularly choice spot, deer may return to the same bedding area time after time.

DAMAGE TO PLANTS. Among the best of signs that deer have in fact visited (and snacked on) your yard are the plants left behind. If you find leaves and twigs snipped off clean, then deer are *not* to blame (at least not this time). Deer have no upper incisors; bottom teeth meet a tough upper pad in the top of the mouth. When a deer has a mouthful, it quickly pulls its head to one side to tear the food free. This leaves a characteristic jagged edge to leaves and torn stems. Rabbits, woodchucks, and the neighbor's pony will

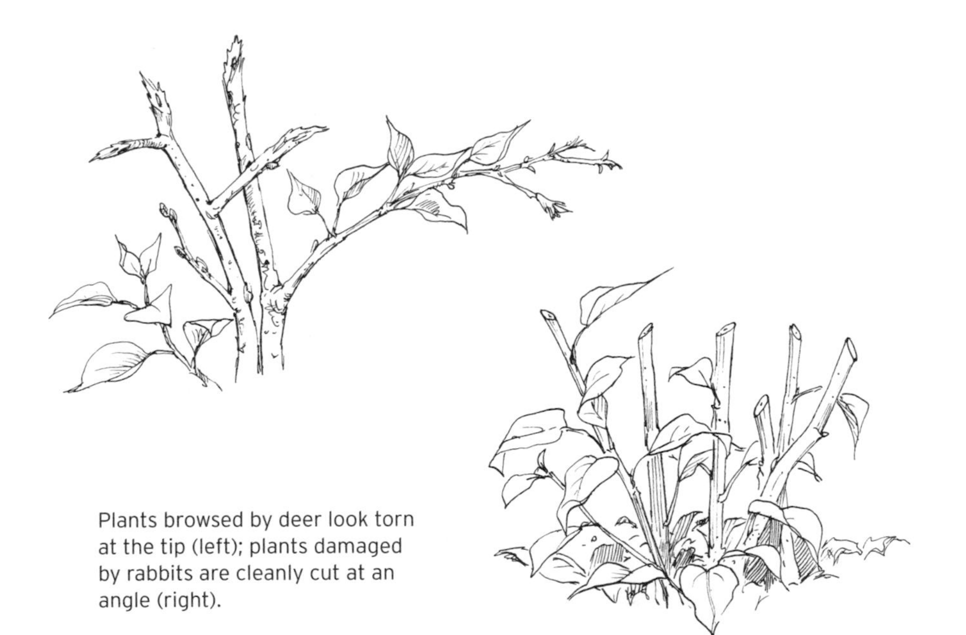

Plants browsed by deer look torn at the tip (left); plants damaged by rabbits are cleanly cut at an angle (right).

bite off plants cleanly. Deer, goats, and llamas must rip them free. In times of famine, deer leave what is known as a browse line — every green stem, bud, or twig of gnawable size gets eaten, from ground level to as high as the deer can reach when standing on their hind legs (usually four to five feet in height, but sometimes as high as seven feet). No bunny could do that, and unless the neighborhood livestock is also starving, it's very unlikely that goats, cows, or llamas will. Deer also like to sample certain vegetables. They take single bites from green tomatoes and gnaw holes in pumpkins and squash, leaving behind ruined, rotting fruit pocked with tooth marks.

Less common, and only in the fall, is the damage left to trees and shrubs when zealous bucks get the itch to scratch. Smooth-barked trees, most commonly with trunks up to four inches in diameter (the larger, older, and more dominant the buck, the larger the tree trunk they pick out), will have the bark shredded from the trunk, about two to three feet off the ground. Trees that are girdled (have the bark removed all the way around the trunk) will die. Just removing a significant amount of bark impairs a tree's vital functions and can lead to distorted growth or a slow death. The ground beneath (and any ground cover) will be stomped and pawed by sharp, pointy hooves. In contrast, porcupine feeding normally results in a littering of bark shavings or scat, not churned-up soil.

Of course, there is no more damning testimony than that of an eyewitness; the best evidence that deer have caused your gardening woes is to catch them in the act. During most of the year, the best time to catch deer in your yard is early morning or evening. Though deer are active on and off for most of the day, they tend to reserve mealtime for the twilight hours. If in doubt, camp out.

Other Injuries and Insults

Not to be outdone by television and politicians, deer have caused problems we could never have imagined until it was too late. In places where rampant deer populations have overburdened the habitat, confrontations with people and deer have reached an all-time high.

Direct Confrontations

Though unusual, deer can attack pets and people (and have done so). Bucks in rut are especially unpredictable and may take out their aggression on anything that they perceive to be a challenge to their territory. At least one attack in Texas left a man trampled and gored to death by a white-tailed buck in rut. Another Texas attack left three surveyors shaken but alive to tell their story. One of the workers was pitched 20 feet into the air by the raging beast.

Traffic Tragedies

Most human suffering attributable to deer is caused by traffic accidents. According to the Insurance Information Institute, between 120 and 150 human fatalities occur annually because of collisions with deer, making deer the number one wildlife killer of man. Although there is no national standard for reporting deer–car collisions, the National Safety Council (NSC) estimates that there are several hundred thousand per year. Most often, the deer does not survive. It is fairly common for the deer to escape the immediate scene; however, the sustained injuries often lead to death from wounds, infection, or predation (injured deer are a favorite target of loose dogs). These collisions also result in approximately 13,000 nonfatal injuries to people and add up to a whopping $1.2 billion in insurance claims for vehicle damage. Deer accidents often go unreported, probably because drivers don't want the accident reported to their insurance company. Considering the

Regional Reported Deer-Vehicle Crash Numbers 2000–2003

State	Est. Pre-Hunt Deer Herd	Reported Deer-Vehicle Crashes*	Deaths	Injuries	Est. Approx. Vehicle Damage Cost**
Michigan	1,800,000	67,760	11	1,193	$115.2 mil
Wisconsin	1,663,000	21,666	13	792	36.8 mil
Minnesota	1,140,000	5,550	5	520	9.4 mil
Illinois	750,000	23,645	2	976	40.2 mil
Iowa	210,000	6,987	2	523	11.9 mil
Totals for Upper Midwest	5,563,000	125,608	33	4,724	213.5 mil

(Courtesy of the Deer-Vehicle Crash Information Clearinghouse (DVCIC) at the University of Wisconsin-Madison)

*It has been estimated that the total number of actual deer-vehicle crashes may be at least twice as large as those reported. In Minnesota, it is believed to be three to four times as large as those reported. As expected, the number of unreported deer-vehicle crashes probably varies from state to state because of different reporting procedures, and few states track the number of carcass collections. The number of reported crashes in Iowa is for all animal collisions.

**Minimum property-damage crash-reporting thresholds can also be different: $1,000 in Iowa, Minnesota, and Wisconsin; $500 in Illinois; $400 in Michigan.

percentage of unreported or uninsured collisions, the real toll could be much higher.

The good news is that, overall, the rate of collisions seems to be diminishing in recent years, most likely, the NSC believes, because motorists have become more aware of the dangers. In 2003 they calculated only 480,000 deer–car collisions, down from 500,000 the year before and a significant drop from 750,000 in 1999. Unfortunately, that doesn't apply to all areas, especially those in which suburban sprawl has continued to encroach on traditional deer territory. The Pennsylvania Department of Transportation (DOT) reported in December 2003 that the incidence of deer–car crashes had risen 22 percent since 1993. With a statewide deer population estimated at more than 1.6 million, the state DOT believes that between 40,000 and 50,000 deer are killed along Pennsylvania's roads every year.

Are You Covered?

It's important to find out whether your insurance policy covers a deer-car collision. Deer accidents don't fall under the standard "collision" portion of your policy. Instead, check your policy for "comprehensive" coverage and note the deductible. This will tell you whether, and how well, you are covered in the event of a deer-car collision. Most insurers don't even factor in a single deer collision when determining premiums.

WHISTLE WHILE YOU WORK. Grill-mounted deer whistles, designed to warn deer of oncoming traffic, have been touted by many, including insurance companies, as a means to prevent collisions with deer. Others have their doubts, because the whistle's effective range is narrow and directly forward — deer can't hear it unless they are standing directly in front of it, in the path of the oncoming vehicle.

Experiments have shown the air-fed whistles to be acoustically ineffective. Because they depend on the flow of air through them to generate sound, the whistles work differently at different vehicle speeds. Whistles mounted on a car driving between 30 and 45 miles per hour produced sounds of either 3 kilohertz (kHz) or 12 kHz. White-tailed deer can hear between 2 and 6 kHz, so they can't even hear the devices that produce 12 kHz sounds. Although deer can hear sounds of 3 kHz, that level of sound is only three decibels different from the road noise of the car, so the sound is drowned out. "All in all, the air-fed whistles do not make sense to me, acoustically," says Peter Scheifle, a researcher in the depart-

Steering Clear

Following these suggestions may help you avoid colliding with deer on the road:

- Heed deer-crossing signs. They are there because at least one deer collision has already happened in that spot. Use extreme caution.
- Drive more carefully during twilight hours, as this is when deer are most active.
- Use high beams after dark and don't overdrive your headlights. Be sure you have time to react to whatever you can see within your limited field of vision.
- Wear your seat belt.
- Exercise more caution during spring and fall — peak times of deer activity.
- Be aware of the surrounding terrain. Remember that deer prefer "edge" habitats. Keep down your speed around such natural deer crossings as golf courses, parks, river or stream crossings, and wooded areas.
- Watch for eyes. In the dark, a deer's eyes reflect lights and will glow at you from the side of the road. Slow down or stop when you spot them.
- If you see a deer, honk the horn to ward it off. Don't flash your head lights; this can cause the proverbial "deer-in-the-headlights" stop and stare.
- If you see a deer near the roadway, slow down and prepare for it to jump right at you, even if it has just crossed from the opposite side.
- If one deer crosses the road in front of you, be prepared for a second. Does usually travel in pairs or small groups.
- If the worst happens and you can't avoid a collision, brake — don't swerve — and let up on the brakes just prior to impact. This allows your vehicle to go over the deer, rather than the deer going through your windshield.
- Although many states allow you to take a deer carcass after a collision, be cautious about touching a downed deer. If it is alive, it may suddenly panic and do you both more damage. The sharp hooves can easily produce serious wounds.

If you do happen to hit a deer, report it to your local game authorities (even anonymously, if you prefer).

ment of animal science at the University of Connecticut. An electronic deer whistle, with a range of 1,500 feet, is currently being tested. Not dependent on the speed of the car, electronic whistles represent the next generation of deer-warning devices.

Even if the deer does hear the sound, how will it react? The goal is to alert the animal to oncoming traffic, so that it is not startled by the sudden approach of a vehicle and has time to react by moving out of the way. But because deer do not instinctively run from strange sounds, the whistle tone may serve only to arouse the deer's curiosity. As a whistle-equipped car approaches, drivers have reported seeing deer standing in the middle of the road looking confused, as if they were trying to figure out what the sound was.

Hi-Tech Tactics

Communities in high crash-risk areas often find dealing with deer populations an overwhelming task that drains resources and creates conflict among community members (especially when prescribed hunts are the topic of debate). Increasing driver visibility by clearing roadside brush or tall grass — in which deer are camouflaged until the instant before they spring into the roadway — can be an expensive proposition for some towns. And keeping roadways clear of carcasses is a job no one enjoys.

Canada is taking a high-tech approach to alerting motorists that deer are near roadways. Infrared technology, designed by NASA to alert fighter pilots to approaching missiles, now monitors a two-kilometer stretch of highway in Kootenai National Park, British Columbia. Infrared cameras rigged to 20-foot-high poles detect heat differences as minute as .01°C — rain or shine, in darkness, through snow or fog, and over a distance of several kilometers. When the system detects an animal, blinking warning signs along the highway warn motorists to slow down. The system costs up to $75,000 to cover one kilometer, but with British Columbia suffering 9,000 wildlife collisions last year, at a cost of $23 million, Canadians figure the system could prove economically feasible. Officials estimate it could cut collisions by as much as 42 percent.

A similar approach is in use in western Washington state. Elk that live near the community of Sequim are outfitted with radio collars. Whenever the elk get within a quarter-mile of the road, the transmitters within the collars activate roadside signs that flash "Warning: Elk X-ing" to alert motorists to their presence.

Although fair warning, visibility, and even deer-population control are great steps towards avoiding a deer collision, nothing beats a thorough understanding of how to drive in deer country. Understanding deer habits, and their consequences for drivers, is your best insurance.

Strange Encounters

Deer can turn up in some pretty unusual places. Reports abound of deer in busy metropolitan areas, stopping traffic and startling the citizenry. One deer that found its way into downtown Washington, D.C., had to be darted and removed.

Deer are phenomenal swimmers and can keep afloat for long periods, especially in winter, when their hollow-haired coats enable them to ride higher in the water. Good thing, too. There are

Breaking In, Breaking Out

Glass is not a concept deer understand. Reports of deer crashing through sliding glass doors are more common than you might think. Once inside, the confused creature predictably panics and tears the place apart.

In one case, near Philadelphia, the homeowner was gone when a four-point buck barged in. Neighbors rallied to rid the house of the intruder. Breaking open another window as an escape route and then opening the door, police and neighbors banged against the sides of the house to shoo out the deer. Suddenly, the deer burst through the broken-out window and charged directly toward the onlookers. Spectators and deer all beat a hasty retreat.

innumerable reports of deer falling into swimming pools and other man-made water hazards, and needing help to get out. The vertical sides of a pool make it impossible for a deer to climb out on its own.

Though no one can prevent all freak accidents, you can at least attempt to protect your property, and the deer. Be sure to fence all pools (eight feet high to deter deer) or cover them when not in use. Make large windows more visible by placing stickers on them or by drawing shades, curtains, or blinds, especially during the times when deer are most active. Never tempt a deer to come indoors — neither of you will enjoy the encounter.

Human Health Risks

No list of deer worries would be complete without mentioning some of the health issues that involve deer. As animals that are constantly on the move in search of forage, deer are often responsible for hosting and translocating ticks that carry diseases that can be spread to humans and other animals. Deer are also hosts for other parasites that may cause *them* no harm but which can endanger livestock or pets.

Lyme Disease

Named for the town of Lyme, Connecticut, Lyme disease was first identified in the United States in the mid-1970s, when physicians observed what they assumed to be arthritis in peculiarly high rates among children from that area. Since that time, Lyme disease has become widespread among the northeastern, mid-Atlantic, and upper-Midwest states, as well as in a small area of northern California. In 2002, the Centers for Disease Control and Prevention (CDC) collected reports of 23,763 cases, a 40 percent increase over 2001. The vast majority of these were from Connecticut, Delaware, Maine, Massachusetts, Maryland, Minnesota, New

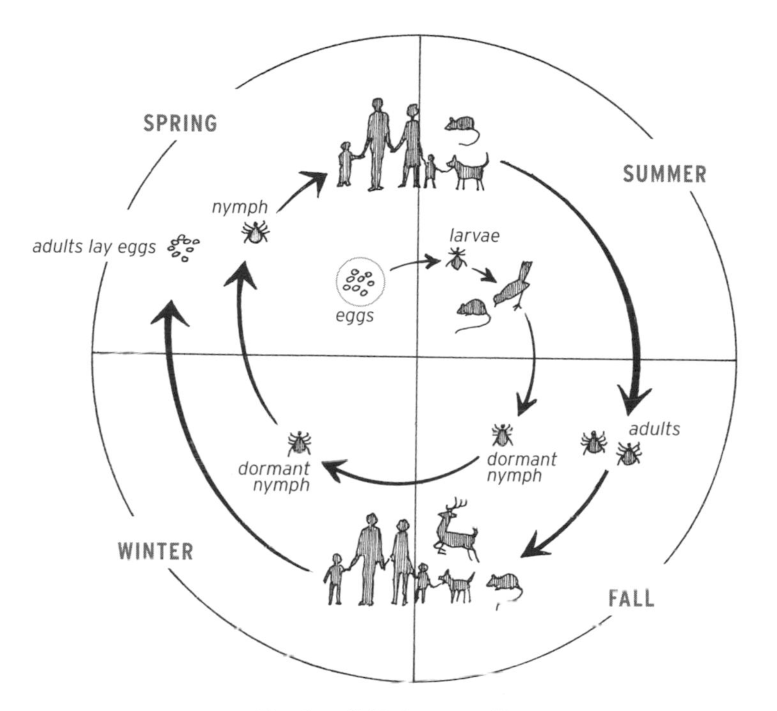

The deer tick's two-year life cycle.

Hampshire, New Jersey, New York, Pennsylvania, Rhode Island, and Wisconsin.

Lyme disease is caused by the bacterium *Borrelia bugdorferi*, which is transmitted by the black-legged tick (*Ixodes scapularis*). In its larval stage, the tick becomes infected with the bacteria while feeding on any number of rodents or other small mammals. In fact, research conducted by the University of Rhode Island and the CDC indicates that deer mice (also called white-footed mice) are the primary reservoir for the bacteria, especially in the East.

Despite being labeled "*deer* ticks," the ticks in question feed on a number of different hosts. They are most likely to be found on deer at the end of their two-year life cycle. In the autumn of their second year, the ticks search for large animals on which to feed and mate, after which the females drop off and lay their eggs in grassy areas or under leaf litter. The tick eggs overwinter, and larvae emerge in the spring to feed on mice and birds, thus repeating the cycle of Lyme-disease transmission.

In order for the bacteria that cause Lyme disease to be transmitted to humans, the infected tick must remain attached for at least 48 hours. Because of this, people are most likely to contract Lyme disease during the spring months, when ticks are in their nymphal stage and are thus small enough (approximately one tenth of a centimeter in size) to feed for long periods without being noticed.

Those who become infected often develop the characteristic "bull's-eye" rash and flulike symptoms — fatigue, chills, fever,

Avoiding Lyme Disease

To prevent being bitten and infected:

- Avoid tick-infested areas, especially May–July
- When outdoors, avoid areas ticks prefer, such as dense brush, leaf litter, and dark, damp areas. Adults are often found on tall shrubs.
- Wear insect repellent containing DEET whenever you venture into deer-tick territory.
- Wear light-colored clothing, so you can spot ticks easily.
- Wear long pants, a hat, and a long-sleeved shirt when possible.
- Wear tall rubber boots for extra protection, as most ticks live at or near ground level.
- When outdoors for long periods, check clothing for ticks every two to three hours.
- If your hair is long tie it back or put it up.
- Large-scale tick control is not yet available, but if Lyme disease or other tick-borne diseases are endemic in your area, consider having your yard treated for ticks. Damminix Lyme Disease Control (a biodegradable product) and other pesticides are effective.
- In addition to, or in lieu of, treating your yard with pesticides, consider raking up leaves, removing tall grass, and clearing brush. These are ideal habitats for tick-bearing rodents, deer, and other wildlife. The leafy debris under your porch or deck is a perfect habitat for rodents and ticks. Keep it clean down there, or your next barbecue party could have uninvited guests!
- Keep woodpiles and other prime rodent-nesting real estate away from your house.

aches, joint pain, and headaches — within 7 to 14 days. Though it is easily cured in the early stages with tetracycline or amoxicillin, not everyone who contracts the disease displays the early-onset symptoms, causing diagnosis to be delayed until advanced symptoms develop and more aggressive treatment becomes necessary. Lyme disease can lead to serious complications if left untreated, such as arthritis, episodes of painful swelling in the large joints; neurological problems, including inflammation of the brain (encephalitis), facial palsy, and nerve inflammation; and, less commonly, heart problems.

Prevention and prompt treatment are your best defenses. Those who work or play in grassy, brushy, or wooded areas — prime habitat for the deer mouse (as well as for deer) — have the greatest risk of being bitten by ticks. Make a habit of checking yourself and others for ticks after every likely exposure. Be sure to wear long sleeves and long pants (and to tuck pant legs into socks, if possible) when you expect to be in a tick-infested area. The CDC has also suggested removing leaf litter from your yard, which may reduce tick populations by more than 70 percent.

Where Ticks Hide — On You

Checking for ticks is not for the squeamish, but it needs to be done regularly during tick season if you go outdoors. Don't be shy to ask for help — you can't see everywhere ticks may lurk. Ticks love the dark, damp creases and folds of your body. Check all along the hairline, behind your ears, in your armpits, behind your knees, and between your legs. Ticks tend to crawl upwards, so if they hopped on at ground level they may migrate as far as your head.

Human Babesiosis

A rarer deer tick–borne disease, human babesiosis (caused by the protozoan *Babesia microti*) is a malaria-like infection that was first diagnosed in the United States on Nantucket Island, Massachusetts, in the 1970s. It can be transmitted along with Lyme disease when a tick feeds — approximately 25 percent of

How to Remove a Tick

First of all, don't panic. Not all ticks carry disease and even if this one does, you stand a good chance of avoiding infection if you remove the beastie within the first 24 hours. To remove a tick once it has become attached to you, firmly grasp it as close to the head as you can with fine-tipped tweezers, and in one steady motion pull the tick up and away from the skin. Be careful not to squish its body.

This next advice is very important! Do not twist it out, cover it with petroleum jelly, douse it with alcohol, or dab at it with a hot match. These ideas aimed at bugging the tick to get it out can actually cause it to inject more mouth secretions into your system, thus increasing the chances of infection. However, don't worry if the mouthparts remain in the skin; they alone do not carry much of the bacteria that cause disease, but rather serve as a conduit to pass them from the tick's body into a host. Put the tick in a jar of alcohol to kill it. Then treat the bite with antiseptic. Write on your calendar the date you were bitten. Keep an eye on the site of the bite and make sure you know what early symptoms of the disease look like, so you can recognize them in case they appear.

reported babesiosis patients simultaneously contract Lyme disease. Although most people infected only with *Babesia* never show symptoms, in rare cases it can be fatal, especially when contracted by people over age 50 or those with weakened immune systems. The disease infects red blood cells, causing anemia. When symptoms are present, they may include lack of appetite, fatigue, fever, sweats, chills, headache, depression, difficulty sleeping, dark urine, and nausea, but not the telltale skin rash of Lyme disease. Symptoms are often mild and develop from one to four weeks after a tick bite. They usually go away without treatment, but in some cases can become severe. Though several hundred cases have been reported in the past 30 years, authorities feel that because most cases are asymptomatic, the actual number of cases could be much higher.

Human Anaplasmosis

Human anaplasmosis (HA) was recognized as a new bacterial disease in 1993, and several hundred cases have since been identified. Until then, it was commonly misdiagnosed as "rashless" Lyme disease because its symptoms are so similar — fever, headaches, muscle aches, chills and shaking, and, less often, loss of appetite, nausea, vomiting, diarrhea, abdominal pain, cough, aching joints, and neurological changes that can alter mental faculties. Like Lyme disease, HA is transmitted by deer ticks. If diagnosed early, HA responds readily to doxycycline or tetracycline.

Pet Projects

Even though deer ticks have preferred hosts — rodents, birds, and deer — they will happily infest your cat, dog, rabbit, or even your horse. Indoor/outdoor pets are handy transport systems, and often deliver ticks right into your home.

To prevent pets from becoming lassie limousines, be sure to treat them regularly with an appropriate tick repellent. There are collars, dips, shampoos, and powders galore for cats and dogs. Once-a-month applications of flea-control products like Advantix or Frontline will also do the trick, and may be the least-toxic option. Be aware that mixing and matching products can overexpose your pet (and anyone who plays with it) to these pesticides. Talk with your veterinarian about which product to use and stick with it.

Be sure to check indoor/outdoor pets for ticks regularly — ideally, every time they come back in the house. Also check their bedding often. This is one reason why pets should have their own bed, separate from yours! Be sure to vacuum carpets and furniture on a regular basis to pick up loose ticks and eggs, and then promptly dispose of (or burn) the bags.

Dogs, cats, and horses can contract Lyme disease and other tick-borne diseases, which makes repellents and tick checks all the more important. Symptoms may include depression, lack of appetite, stiff or sore (arthritic) joints or other sudden pain, limping, fever, or cough. Your vet can perform a blood test to confirm whether or not your pet has been infected, and can provide advice on available treatments.

Other Threats to Man and Beast

Other health concerns involve disease agents that can be transmitted to livestock. Among the most worrisome is the spinal meningeal worm, which is carried by whitetails. It is fatal to mule deer, blacktails, elk, moose, llamas, alpacas, goats, and some sheep. In addition, deer flukes, brucellosis, tuberculosis, and anthrax can all be carried and transmitted by deer. And chronic wasting disease, a newly identified illness, has already had devastating outbreaks.

Spinal Meningeal Worm

The spinal meningeal worm (*Parelaphostrongylus tenuis*) is a nematode parasite, sometimes known as the brainworm and also deer worm or deer meningeal worm (the white-tailed deer is its natural host). Though not common, it is most prevalent around the Great Lakes. The parasite passes from the deer (through feces) to snails or slugs, then returns to deer or to other animals when they accidentally consume the snails or slugs. Infected deer rarely show symptoms, but other animals — especially llamas and alpacas, and sometimes goats and sheep — can be debilitated by the disease, in which the larvae of the worm migrate through the nervous system on a pilgrimage to the brain. A mild infestation localized near the limbs may produce only a slight limp, whereas a severe infestation that invades the spinal cord can leave an otherwise bright and alert animal paralyzed. Because all hosts except the white-tailed deer are "dead-end" hosts that can't perpetuate the life cycle of the parasite, they cannot pass on the disease to each other or to other types of livestock.

Livestock infection is best prevented by regular deworming with Ivermectin and by pasturing susceptible animals away from deer. Keeping pens and pastures dry, clean, and thus less attractive to snails and slugs is also recommended.

Deer Flukes

The deer fluke (*Fascioloides magna*) is a parasitic flatworm that infests the liver of a deer, where it can release up to 4,000 eggs per day. Although deer can harbor significant numbers of flukes with no apparent symptoms, cattle and sheep are not so lucky. Nor are moose. A single fluke can kill a sheep, whereas cattle and moose display symptoms typical of parasitic infestation, such as weight loss, anemia, and edema. The deer fluke life cycle also depends on snails and slugs. Cercaria (a tadpolelike form of the fluke) are shed by snails and encyst themselves on plants that are later eaten by deer or livestock. Because livestock are not the natural hosts of the fluke, their livers do not support them as those of deer do, and the parasite cannot reproduce.

Brucellosis

Brucellosis is a disease caused by *Brucella* bacteria. In addition to deer and elk, cattle, camelids (llamas and alpacas), sheep, and goats can become infected. Humans can become infected with *Brucella* through inhalation or by handling infected animals or animal products (meat, milk, cheese). There are only 100 to 200 human cases of brucellosis reported in the United States each year; it is more common in countries without public or animal health programs.

Tuberculosis

Caused by the bacterium *Mycobacterium tuberculosis,* tuberculosis (TB) was once the leading cause of human death in the United States. TB is also commonly associated with deer, elk, and bison. Both captive and free-ranging deer in Michigan have been reported to be infected with *M. bovis,* the same strain that was nearly eliminated from domestic cattle since concentrated efforts began in 1917. Both *M. tuberculosis* and *M. bovis* can spread to humans, most likely through contact with infected livestock. The concern is that deer transmit the disease to cattle by browsing on feed and leaving traces of bacteria-laden saliva that the cattle later consume. Research conducted in 1997 at the National Animal Disease Center in Ames, Iowa, supported this idea, indicating that deer can transmit *M. bovis* to each other and to cattle through nasal secretions, saliva, and, less frequently, urine and feces. Because there is no test for TB in deer meat, hunters are advised to wear gloves when dressing out deer, and to thoroughly cook any meat that may be suspect for TB.

Anthrax

More than just a vehicle for terrorism, anthrax is a naturally occurring disease that infects deer, livestock, and humans. Caused by a specialized type of bacterium (*Bacillus anthracis*) that produces spores, it is found sporadically throughout the United States, primarily along the old cattle trails between Texas and Canada, where spores from dead cattle settled in the soil. The spores can survive for many years in the soil (or in water) — according to one extension service report, 100 years or more. Lying dormant beneath the soil, the bacteria "bloom" after periods of wet, cool weather followed by a dry hot spell. As the spores hatch and multiply, they rise to the surface and contaminate the soil, grass, weeds, and any low-growing forage. Deer, cows, goats, sheep, and other animals then consume the spores while they feed or drink, or absorb them through open wounds.

There are three possible modes of infection in people — inhalation, gastrointestinal (by eating spores), and cutaneous (through the skin). Worldwide, cutaneous infection from handling infected livestock — or products from infected animals — accounts for 95 percent of the cases of human anthrax. Here, the most probable means of infection is handling dead anthrax victims inappropriately or eating undercooked infected meat. Treated cases of cutaneous anthrax have less than a 1 percent mortality rate. Early treatment is essential for recovery; penicillin, doxycycline, and ciproflaxin are all effective.

Chronic Wasting Disease

Chronic wasting disease (CWD) in deer and elk is one of a group of diseases, known as transmissible spongiform encephalopathies (TSE), that attack the nervous system and brain. These include bovine spongiform encephalopathy (BSE), commonly called mad cow disease, and Creutzfeldt-Jakob disease (CJD) in humans.

By the 1990s, when alarming reports of herds of emaciated deer hit the news, CWD had already been around for decades, perhaps even longer. The first case identified as CWD was in a captive mule deer in Colorado during the late 1960s. By 2001, it had spread as far away as Wisconsin. Natural migrations of deer most likely caused the gradual spread throughout the original area, but researchers figure that deer being transferred among game farms probably account for the disease jumping great distances. It is now assumed that the disease passes back and forth between wild and captive populations. At this time, researchers don't know exactly how CWD is transmitted from one animal to the next, only that it is believed to be highly contagious.

Like all forms of transmissible spongiform encephalopathy, CWD progressively attacks the nervous and lymphatic systems. As the brain degenerates, the animal's behavior changes dramatically. Affected animals become listless, avoid other animals, stand or walk with the head lowered, and have a blank, absent expression. Some walk around repeatedly in set patterns. However, CWD is most readily recognized as a fatal wasting syndrome, characterized by reduced appetite leading to emaciation and, always, ultimately death. Most animals perish within several months of the onset of symptoms; rarely, an animal may last a year or more. Once infected, there is no treatment or cure, and as of this writing, there is no vaccine for CWD.

It's not certain whether CWD can spread to livestock; so far, domestic livestock seem resistant to CWD infection. In controlled settings, sheep, cattle, and goats living with infected deer have not contracted the disease, and one experiment that fed cattle CWD-infected feed for more than six years failed to produce the disease in the cattle. (This is one difference between CWD and its relative BSE, which is spread through contaminated feed.)

Perhaps the biggest scare in the spread of CWD is a suspected link to human cases of Creutzfeldt-Jakob disease (CJD), especially considering the cases of variant CJD that were linked to an outbreak of mad cow disease in Great Britain. Several fatal cases of the disease, initially suspected to have links to consumption of venison, have been reported in the United States since the 1990s. To date, however, only two cases of CJD have been linked to venison from CWD-endemic areas.

It is possible that not enough people have been exposed to CWD to determine to what degree humans are susceptible. Due to the low incidence of exposure, the studies into possible links have been limited, but as the disease spreads and more people are exposed, more studies will be forthcoming.

Even with the risk to humans considered low, officials recommend avoiding exposure to CWD-infected animals until conclusive evidence rules out CWD as a potential public health hazard. Hunters are advised not to eat meat from thin, sickly deer or elk, or from any that test positive for CWD, and not to ingest tissue in which the infectious agent is known to settle (such as the brain, spinal cord, eyes, tonsils, lymph nodes, or spleen).

In addition, wildlife managers recommend not feeding deer or elk, and not setting out salt blocks, as that concentrates populations, thereby increasing the opportunities for contagious infections like CWD to spread. Sick or dead deer or elk (tame, farmed or wild) should be reported to your local Division of Wildlife or Department of Fish and Game.

FOUR

Deer-o-Scaping

LET'S HOPE that the first three chapters have helped increase your understanding of deer and how they live. Perhaps you've reached a vantage point from which these wild creatures merit newfound respect. Now we'll apply this knowledge to limit deer damage to your yard and garden. The first option involves reconsidering which plants you select for your property.

Just as xeriscaping was developed to meet the demands of gardening with limited water, so too can the gardener establish and maintain her plot under a plan to prevent deer damage. Deer-o-scaping can be defined as the style of gardening that incorporates techniques designed to discourage deer damage. It depends on three things:

- Avoiding the use of plants deer love
- Choosing plants you love and deer don't
- Combining deer-resistant plants and garden designs to create a landscape you'll love and they won't

Why Your Yard May Lure Deer

Very different gardens in very different settings draw deer. Deer in arid regions may flock to your garden because it is irrigated. During the summer, our friends who live on a high, dry hill often lose their well-watered plants to sneak thieves in the night. Are your roses, raspberries, or crab apples a target? Even deer in lush rural areas may make a special trip to your yard to sample delicacies that they just can't find elsewhere.

And it's not just about food. What the deer find so irresistible may have as much to do with the surrounding area as the individual yard. Deer in developed areas may simply find quiet shelter in your yard that the neighbor's pool and tennis court don't offer. Although there may be little you can do to change your setting, understanding the environment it provides can be very helpful in understanding what deer find so appealing about your particular plot.

Rural Garden Parties

We used to automatically associate deer with a rural setting, and though they have certainly spread beyond the countryside, deer are still most common in rural areas. The most obvious reason for this is that the habitat is still more "deer friendly" in the country — a variety of native plants on which to browse, streams or livestock troughs for water, natural cover to hide them during the day, and fewer gardeners shooing them away.

Large woods that break into open expanses of pastures or farm fields, the edge habitat preferred by deer for millennia, are still more common in rural areas than in suburban landscapes. Some regions are blessed with more wooded, edge habitat than others — and more deer. The lush woods of Wisconsin support anywhere from 68 to 115 deer per square mile, compared with the 2 to 5 deer per square mile averaged across Wyoming, Nebraska, and Colorado. The varied landscapes and swamps of southern Alabama boast nearly twice the deer per square mile as northern Alabama pine forests. Ideal deer habitat — a mix of conifer and deciduous trees with open meadows — along with nearly nonexistent predation simply maintains more deer and/or elk than do arid plateaus, open plains, piney woods, or your neighborhood park.

Rural gardens — those pastoral oases between wheat fields, woods, and cow pastures — attract deer because they offer a neat little buffet right in the middle of the deer's domain. Rather than search their entire range for a suitable mix of vegetation, they find everything they could ask for in a single convenient location. It's like one-stop shopping for deer. And nearby, deer can usually find cover, water, and possibly other enticements, such as feed, salt, and mineral blocks for livestock. Deer do not fear livestock, and can frequently be seen browsing in pastures next to grazing cows, horses, goats, and other domestic animals. They may even be comforted by their presence. It's almost as if they think, "Gee, if these guys think it's safe to hang out here, it must be okay for us, too." A rural landscape also offers less commotion, noise, and olfactory distraction, so deer naturally feel more at home.

Rural gardens attract deer because they offer a neat little buffet right in the middle of the deer's domain. It's like one-stop shopping for deer.

If your garden is in a rural setting, first check the woods at the edge of your landscape for deer trails, then check nearby pastures and woods for trails. Whitetails, especially, are creatures of habit and will use the same trails over and over again.

Suburban Splendor

Suburban growth continues to encroach upon deer country. Newly developed 'burbs disorient deer, which have probably used the same trails to the same food sources and water for generations. If your yard is in an area that was recently developed — from woods, farm fields, or other seemingly "unused" land — don't be surprised if you find it is right smack dab in the middle of where resident deer expect to find their dinner. Even more-established suburban areas, those lovely enclaves of well-kept yards and mature trees, may still be considered "home sweet home" to deer whose ancestors claimed "your" land generations ago. The deer find themselves suddenly displaced and exposed to traffic, noise, strange fumes, and dogs. Habitat areas are reduced and fragmented, which increases the deer pressure on remaining resources. With less natural habitat and food available, deer are lured into yards that are well stocked with new plants. Eventually, these plants may even replace the deer's normal diet.

Suburban gardens may offer deer the same food choices as rural plots, but in this setting there may be dozens of gardens within the range of each deer. If your garden also provides, or is near, brush or bushy cover, it may become a favored stopover along the deer's garden tour. Culs-de-sac and dead-end streets, in particular, often offer just the right combination of quiet, cover, and cuisine.

Urban Invitations

Despite everything that deer hold dear — bucolic surroundings, peace and quiet, edge habitat, and nearby water — they have increasingly become a problem in urban settings. Every year, more and more deer adapt to life in cities, within leaping distance of rush-hour traffic and a nibble away from municipal hedges. Here the yards are smaller, but public parks and other open areas still provide the basics of food, water, and cover, and the only predator is on four wheels. Urban infiltration occurs most often in spring, when young deer are chased away by their mothers in preparation for birthing new fawns. If these confused and inexperienced young deer survive their initial forays into strange territory and discover the rewards offered by community gardens and well-fertilized park lawns, they may return.

Deer don't tend to have difficulty adjusting to city life; those that find their way downtown, after all, are not traveling down from the mountains, but rather commuting in from the suburbs. They will have already learned to dismiss the sights, sounds, and smells of the city. As yearling deer establish their own territories, they tend to stay close to their mother's home range, but with so many deer already inhabiting surrounding areas, each succeeding generation inches closer to urban settings.

Often hemmed in by highways, housing developments, and other human barriers, city deer face real challenges in finding enough to eat and a place to hide. Virtually any greenery in the city becomes a meal, and the sparsest shrubbery can provide sufficient cover. Parks often host deer until damage becomes evident. In this context, any lush, well-fertilized plants and lawns are beacons, and your city garden, an oasis.

Deer Favorites: What Deer Want from a Plant

The first rule of deer-o-scaping is to avoid plants that deer actively seek out. Unfortunately, this varies by area, time of year, species, and even individual deer. Previous habits, the taste of plants, weather conditions, and access to other foods also play a role in which foods deer will prefer. Deer food preferences are not static. It's important to realize that there is no one-size-fits-all answer to what deer prefer; some plants, however, are prime targets just about anywhere you find them. Like us, deer go for certain foods because — to them, at least — they taste good.

A deer's seasonal food preferences have a lot to do with what else is available. Newly emerging crocuses in early spring, tender and tasty, and succulent tulips a little later on, will draw deer because there is precious little else out there to eat at that time of year. This brings up another important point: deer damage is often seasonal. Deer are most likely to risk dining in your yard or garden during early spring, when natural food sources are scarce, and again during late summer and fall, when they instinctively fatten up for the coming winter. In early summer, when native flora is at its peak, you just might get a break from the hungry horde. Plump, watered pansies taste good any time of year, but surely never better than in late summer, when everything else is dried out. Late summer and early fall bring out the munchies in deer, which may now accost foods they ignored throughout the growing season. Wintertime sees dormant buds and twigs from fruit trees and ornamental shrubs, along with evergreens becoming deer fodder.

Why deer seek out particular foods is not always clear. Sometimes it's based on nutritional needs; at others it's the same reason we seek out chocolate. Some plants, such as soybeans, or any well-fertilized greenery, appeal to a deer's need for protein, especially during the spring and summer, as does nurse young, bucks tend to their antlers, and last year's babies grow and mature. The nitrogen necessary for forming proteins can be stored in the body only in small amounts, so a constant supply is needed during periods of growth and peak demands. Because deer meet about one third of their water requirements through moisture derived from food, another attraction is high water content.

Deer love tender plant parts that are succulent with water. New growth — buds, shoots, and tender leaves — are deer delicacies. They always go for the outside growth of a plant first, as this is where the newest, most tender growth occurs. It's also easiest to nibble on the move by just trimming the edges. Certain aromas can lure deer closer, the way Grandma's apple pie cooling on a window ledge drew us to the kitchen when we were young. And sometimes it just seems like deer go for the latest additions to the landscape. Just in case, don't ever leave price tags on new plants — especially expensive ones.

Factors Affecting Food Selection

Although some people think that hungry deer will eat anything, the truth is more complicated than that. Deer are selective feeders and choose plants, and the parts of the plants they eat, based largely on meeting their nutritional needs. Whether these choices are instinctive, learned, or a combination of the two isn't completely understood. What is known is that even during hard times, deer are quite accomplished at finding food that provides enough nutrition to sustain them. Deer may consume a wide variety of plants, but there are some common denominators in their selections. Deer prefer plants and plant parts that are high in protein and energy (carbohydrates). They also gravitate to plants that are rich in minerals and salts, like recently fertilized garden plants.

As anyone who has ever landscaped with deer-favored plants can attest, deer will make pigs of themselves when they find their favorite foods. And they will do it over and over again. A friend of mine who loves roses used to have several well-tended bushes along the walkway of her woodland home. Not anymore. Once the deer chanced upon the poor rosebushes, they devoured the plants. To make matters worse, the deer soon discovered other plants and the vegetable garden. Now my friend seeks out plants that deer won't eat.

The Mystery of the Lists

If you look through enough deer plant lists, you'll discover that some plants appearing on preferred-food lists show up on deer-resistant ones, and vice versa. Examples of these are clematis, irises, forsythia, dahlias, vinca, trilliums, and peonies. Some lists even claim tulips and hostas as "deer resistant." What gives?

It's easy to wonder if the list makers are full of fertilizer, but even lists compiled by very reputable sources disagree on particulars at times. Blame it on regional and individual preferences. Regional preferences often depend on availability factors. For example, in parts of Oregon, deer don't bother tulips. That's likely because the gardening climate of Oregon offers so many other food options at the time tulips emerge. So gardeners, researchers, and extension agents conclude that deer avoid them. Elsewhere, however, tulips are among the first garden flowers to go. Another difference is which varieties of specific plants grow in which areas. In short, it's important to realize just how variable deer preferences can be and why.

Deer Favorites

Even though it may seem that deer will eat anything that holds still long enough, the fact is that they have distinct preferences. Gardeners, beware: plants on the list below should, like double butter-brickle ice cream, be indulged in sparingly. These plants tempt deer into your garden when they might otherwise never set hoof inside the gate. Keep in mind that, like deer-resistant plants, deer preferences vary from one region to another.

TREES, SHRUBS, VINES

Azaleas (*Rhododendron* spp.): both deciduous and evergreen types

Apples (*Malus* spp.): fruit and twigs

Atlantic white cedar (*Chamaecyparis thyoides*): needles, shoots

Camellia (*Camellia* spp.): both young and mature plants, leaves, and blooms

Cherries (*Prunus* spp.): sweet, tempting fruit often grown on dwarf trees

Cornelian dogwood (*Cornus mas*): new blooms or leaves in the spring; trunks are a prime target for rubbing bucks in fall

Eastern White Pine (*Pinus strobus*): the smooth bark of young trees and twigs in winter

English/American hybrid yew (*Taxus baccata*): Yew slows the heart, and one pound of it will kill a horse, yet deer eat yew with no problems.

English ivy (*Hedera helix*): another moisture-rich, leafy favorite, heavily planted and readily available in many areas

Euonymus (*Euonymus alatus*): foliage and shoots in spring and summer; small twigs and bark in winter

Fir (*Abies* spp.): a major winter food source for moose, and commonly browsed by whitetails

Fringe tree (*Chionanthus virginicus*): buds and tender leaves; twigs in winter

Hybrid tea roses (*Rosa odorata* hybrids): petals and tenderest foliage

Hydrangeas, mophead (*Hydrangea macrophylla*) and **oakleaf** (*H. quercifolia*)

Japanese yew (*Taxus cuspidata*): foliage is nibbled with gusto

Mountain ash (*Sorbus* spp.): fruit and shoots; related to the apple

Peach (*Prunus persica*): tender leaves and tips of twigs

Plum (*Prunus* spp.): buds and ripened fruit

Pear (*Pyrus* spp.): buds and fruit; twigs in winter

Saucer magnolia (*Magnolia soulangeana*): flowers, buds, and twigs in winter

Western yew (*Taxus brevifolia*): tender needles, buds; twigs in winter

VEGETABLE GARDEN PLANTS

Beans (*Phaseolus* spp.): tender new seedlings go first (high protein and moisture content)

Blackberries and raspberries (*Rubus* spp.): lush leaves and delectable fruit

Broccoli, cauliflower (*Brassica* spp.): young tender leaves and shoots later in the season

Grapes (*Vitis* spp.): all parts

Lettuce (*Lactuca* spp.): tender, tasty, and high in water content

Peas (*Pisum satirum*): all parts; high in protein

Raspberries (*Rubus* spp.): leaves and fruit

Strawberries (*Fragaria* spp.): tender leaves and berries

Sweet corn (*Zea mays*): young tender stalks and sugar-filled ears

HERBACEOUS ORNAMENTALS

Dahlias (*Dahlia* spp.): new shoots and young leaves along the stalks

Daylilies (*Hemerocallis* spp.): flower buds and blooms

Hostas (*Hosta* spp.): fleshy, high-moisture leaves

Impatiens (*Impatiens* spp.): large, succulent blooms, especially those of New Guinea impatiens

Pansies (*Viola* x *wittrockiana*): all parts (usually planted when deer are at their hungriest — in early spring and early fall)

Phlox (*Phlox* spp.): leaves, buds, and sometimes flowers

Spring bulbs (various): **Crocus** (*Crocus* spp.), **Grape hyacinth** (*Muscari* spp.), **Tulips** (*Tulipa* spp.), and other early, tender shoots and blooms

Trilliums (*Trillium* spp.): blossoms, leaves, and tender stalk. Consumption likely depends on its early debut in the woods, garden, or border and on what other food options are available

An Expanded Palate

Mule deer add a twist to the list by seeking out such plants as junipers (*Juniperus* spp.), elkweed (*Frasera* spp.), forsythia (*Forsythia* spp.), firethorn (*Pyracantha* spp.), viburnum (*Viburnum* spp.), bugleweed (*Ajuga* spp.), pinks (*Dianthus* spp.), lavender (*Lavandula* spp.), wormwood (*Artemisia* spp.), cotoneaster (*Cotoneaster* spp.), lilac (*Syringa* spp.), snapdragons (*Antirrhinum* spp.), cinquefoil (*Potentilla* spp.), and barberry (*Berberis* spp.), most of which are plants deer normally avoid.

Plants Deer Avoid Eating

Nature equipped deer, like other animals, with an innate sense of what is and what isn't good for them. They will munch edible mushrooms contentedly but never nibble unsafe varieties. They may browse your browallia but forsake the foxglove. When it comes to poisonous plants, they just know. However, some plants they avoid are more like spinach on a toddler's plate — they just don't *like* them. Four things dictate which plants deer usually will not eat:

- The deer's previous encounter with the plant. If a plant makes them sick or is associated with a bad experience, deer tend to avoid it in the future.
- The degree of hunger. If there's nothing else to eat, suddenly spinach doesn't taste so bad.
- The individual deer. (Hey, some kids *like* spinach.)
- General deer distaste. Deer often avoid specific types of plants, such as those with a strong aroma, fuzzy or prickly texture, or bitter or alkaloid taste.

"Plants deer won't eat" is, of course, a relative and ever-changing category. If a deer tries an unpalatable food in the midst of plenty, it will be less likely to go for seconds. Most plant foods reach their prime nutritional content in spring and early summer, becoming less tasty, tougher, and overall less desirable as they mature. However, if a normally rejected food constituted the only sustenance when starvation threatened, a deer will remember it as a good food.

Deer in need of food, such as starving deer and lactating does, grow bolder. Other instincts and fears become clouded by the need for food. They will disregard all sorts of deterrents, usually in winter when snow cover and the natural dormancy of plants make other foods unavailable, and bravely test previously untouched gardens and plants. Starvation, though, is not confined to wintertime in areas where deer overwhelm the habitat. If the

Plants That (Usually) Repel Deer

Some plants actually deter deer, especially those that are highly fragrant. Heavy scent masks other odors, virtually jamming the deer's predator-alert sensors and making them uneasy. The most effective use of highly aromatic plants to deter deer is to plant them in combination. A confusing array of heavy scents is difficult for deer to sort through. There is no guarantee as to how individual deer will react, but based on research and the experience of thousands of gardeners, the following plants appear to be deer bane.

Common Name	Botanic Name	USDA Zone	Soil	Light
Catmint, catnip	*Nepeta* spp.	3-8	Well-drained	Sun
Chives, garlic, onions	*Allium* spp.	5-10	Loose	Sun
Honeybush	*Melianthus major*	8-10	Acid	Partial shade
Lavender	*Lavandula* spp.	5-9	Well-drained	Sun
Mint	*Mentha* spp.	4-9	Any	Any
Sage	*Salvia officinalis*	4-8	Well-drained	Sun
Society garlic	*Tulbaghia violacea*	8-10	Fertile	Sun
Thyme	*Thymus* spp.	4-10	Well-drained	Any

deer density in your area is high enough, if drought or a killing frost affects forage, or if prime food sources such as acorns are in short supply, you may encounter starving deer anytime.

To complicate matters further, deer are inconsistent when it comes to diet. They have preferences, but no absolutes. Except for what is downright toxic, deer don't seem to agree much on what they like and dislike, or on how much they like or dislike certain plants. The menu changes from deer to deer, year to year, and even season to season among the same deer. As gardeners, we have to work from generalities, tempered by our individual experiences.

Deer do find several things unappetizing. They usually don't eat plants with a coarse, fuzzy, bristly, or spiny texture. They generally don't like leaves or stems with milky sap. They also shun plants with an intense scent. Some plants just taste yucky — deer have been observed taking a "test taste" and virtually spitting out offending plants. For instance, deer don't like plants that are high in tannins because the bitter taste normally warns them of toxins. The bitter taste of alkaloids also alerts deer to plants that may be toxic. Even visual cues can deter deer from plants they have learned can make them sick, such as the deep purple fruit of deadly nightshade (*Solanum* spp.).

Getting Down to Specifics

The lists on the following pages identify plants that are least likely to be consumed or seriously damaged by deer. The lists are a guide, not a guarantee, however. Deer preferences vary from region to region; many plants not eaten in suburban Chicago can be deer fodder outside Dallas. Use the lists as a way to *start* thinking about what to plant in your own garden.

You should consider the listed trees as deer resistant after the trees reach a reasonable level of maturity. Curious nibbling may seriously injure very young trees of almost any kind. Protect young plants until they outgrow the danger.

As you plan your garden, consider carefully that as the seasons change, so do the dietary habits of deer. Because deer feed most actively in early spring and early fall, choose the most resistant varieties possible for these times of year. Plants that emerge or bloom in summer are often less susceptible, as deer have other choices, so gambling on a few less-resistant choices may pay off in colorful rewards. But let your experience be your guide. If you have high deer pressure in summer, stick to the more resistant varieties. Winter damage is greatest to trees and shrubs with appetizing bark and twigs. Mature conifers often fall victim to browse during winter, but usually rebound well. Try to keep a far-sighted eye on the course of the year in your yard, and you will enjoy a wider range of interesting, deer-resistant plants.

Bee balm (*Monarda* spp.)
Bergenia (*Bergenia* spp.)
Black-eyed Susan (*Rudbeckia hirta*)
Butterfly weed (*Asclepias tuberosa*)
Columbine (*Aquilegia* spp.)
Coreopsis (*Coreopsis* spp.)
Cranesbill (*Geranium* spp.)
Fleabane daisy (*Erigeron* x *hybridus*)
Foam flower (*Tiarella cordifolia*)
Gentian (*Gentiana* spp.)
Geum (*Geum* spp.)
Goldenrod (*Solidago* spp.)
Hellebore (*Helleborus* spp.)
Hens and chicks (*Sempervivum* spp.)
Hibiscus (*Hibiscus* spp.)
Iris (*Iris* spp.)
Jacob's ladder (*Polemonium caeruleum*)
Marsh marigold (*Caltha palustris*)
Meadow rue (*Thalictrum* spp.)
Meadowsweet (*Filipendula* spp.)
Monkshood (*Aconitum* spp.)
Peony (*Paeonia* spp.)
Phlox (*Phlox divaricata*)
Pinks (*Dianthus* spp.)
Purple coneflower (*Echinacea purpurea*)
Rock cress (*Arabis caucasica*)
Rose campion (*Lychnis coronaria*)
Russian sage (*Perovskia atriplicifolia*)
Salvia (*Salvia* spp.)
Sedum (*Sedum* spp.)
Shasta daisy (*Chrysanthemum* spp.)
Snakeroot (*Eupatorium rugosum*)
Sneezeweed (*Helenium autumnale*)
Snow-in-summer (*Cerastium tomentosum*)
Speedwell (*Veronica* spp.)
Toadflax (*Linaria* spp.)
Valerian (*Valeriana officinalis*)
Violet (*Viola* spp.)
Yarrow (*Achillea* spp.)

HERBS

Artemisia (*Artemisia absinthum*)
Basil (*Ocimum basilicum*)
Borage (*Borago officinalis*)
Catmint (*Nepeta* x *faassenii*)
Chamomile (*Matricaria* spp.)
Chives (*Allium shoenoprasum*)
Comfrey (*Symphytum* x *rubrum*)
Dill (*Anethum graveolens*)
Fennel (*Foeniculum vulgare*)
Feverfew (*Tanacetum parthenium*)
Garden sage (*Salvia officinalis*)
Germander (*Teucrium chamaedrys*)
Hyssop (*Hyssopus officinalis*)
Lamb's ear (*Stachys byzantina*)
Lavender (*Lavandula angustifolia*)
Lemon balm (*Melissa officinalis*)
Mint (*Mentha* spp.)
Mullein (*Verbascum* spp.)
Oregano (*Origanum vulgare*)
Parsley (*Petroselinum* spp.)
Rosemary (*Rosmarinus officinalis*)
Rue (*Ruta graveolens*)
Savory (*Satureja montana*)
Tansy (*Tanacetum coccineum*)
Thyme (*Thymus* spp.)

Northeast

Brad Roeller is the manager of display gardens at the Institute of Ecosystem Studies in Millbrook, New York (USDA Zone 5), and for the past 10 years, he has been testing landscape plants for deer resistance. The plants listed below are those he's found to have the best resistance to browsing in his area.

DECIDUOUS TREES

Beech (*Fagus* spp.)*
Birch (*Betula* spp.)*
Black locust (*Robinia pseudoacacia*)*
Maple (most *Acer* spp.)*
Oak (most *Quercus* spp.)*
Thornless honeylocust (*Gleditsia triacanthos* var. *inermis*)*

EVERGREEN TREES

California incense cedar (*Calocedrus decurrens*)*
Spruce (most *Picea* spp.)
Western and Oriental arborvitae (*Thuja plicata* and *T. orientalis*)*

DECIDUOUS AND EVERGREEN SHRUBS

Barberry (*Berberis* spp.)**
Bayberry (*Myrica pennsylvanica*)*
Beautybush (*Kolkwitzia amabilis*)
Boxwood (*Buxus* spp.)
Bush cinquefoil (*Potentilla fruticosa*)
Japanese and mountain pieris (*Pieris japonica* and *P. floribunda*)
Lilac (most *Syringa* spp.)*
Magnolia (most *Magnolia* spp.)*
Pfitzer juniper (*Juniperus pfitzeriana*)
Red buckeye (*Aesculus pavia*)*
Spirea (many *Spiraea* spp. including *S.* x *bumalda* cultivars, *S. japonica*, *S. nipponica* cultivars, *S.* x *vanhouttei*, *S. dolchica*, and *S. prunifolia*)
Summersweet (*Clethra alnifolia*)*
Viburnum (many *Viburnum* spp., and their cultivars, including *V. acerifolium*, *V. alnifolium*, *V.* x *burkwoodii*, *V.* x *carlcephalum*, *V. carlesii*, *V. cassinoides*, *V. dentatum*, *V. dilatatum*, *V. lantana*, *V. lentago*, *V. opulus*, *V. plicatum* var. *tomentosum* cultivars, *V. prunifolium*, *V.* x *rhytidophylloides*, *V. rhytidophyllum*, *V. sargentii*, and *V. trilobum*)

PERENNIALS

Aaron's beard (*Hypericum calycinum*)†
Alum root (*Heuchera* spp.)
Anemone (*Anemone* spp.)†
Archangel (*Angelica* spp.)†
Astilbe (*Astilbe* spp.)
Avens (*Geum* spp.)
Baby's breath (*Gypsophila paniculata*, *G. repens*)

*May be selected for food during times of harsh winters and high deer populations, when deer are approaching starvation.
**Avoid the popular Japanese barberry as it is an extremely invasive exotic which displaces many indigenous species. Utilize the William Penn barberry (*B.* x *gladwynensis* 'William Penn'), as it appears to be sterile.

Balloon flower (*Platycodon grandiflorus*)
Barrenwort (*Epimedium* spp.)†
Basket of gold alyssum (*Aurinia saxatilis*)
Beardtongue (*Penstemon* spp.)
Bear's breeches (*Acanthus* spp.)†
Bee balm (*Monarda* spp.)†
Bergenia (*Bergenia* spp.)†
Big-root geranium (*Geranium macrorrhizum*)†
Blanketflower (*Gaillardia aristata, G.* x *grandiflora*)†
Blazing star (*Liatris* spp.)†
Bleeding heart (*Dicentra* spp.)†
Bloodroot (*Sanguinaria canadensis*)†
Blue star (*Amsonia tabernaemontana*)†
Blue-eyed grass (*Sisyrinchium angustifolium*)†
Boltonia (*Boltonia asteroides*)
Bugbane (*Cimicifuga* spp.)
Bugleweed (*Ajuga repens*)†
Buttercup (*Ranunculus* spp.)†
Campanula (*C. carpatica, C. rotundifolia*)
Campion (*Lychnis* spp.)
Candytuft (*Iberis sempervirens*)
Catmint (*Nepeta* spp.)†
Celandine poppy (*Stylophorum diphyllum*)†
Chrysanthemum (*Chrysanthemum* spp.)
Cinquefoil (*Potentilla* spp.)†
Columbine (*Aquilegia* spp.)
Comfrey (*Symphytum* spp.)†
Common globeflower (*Trollius europaeus*)†
Common tansy (*Tanacetum vulgare*)†
Coneflower (*Rudbeckia* spp.)
Culver's root (*Veronicastrum virginicum*)†
Dame's rocket (*Hesperis matronalis*)
Dead nettle (*Lamium* spp.)†
English lavender (*Lavandula angustifolia*)†
Evening primrose (*Oenothera* spp.)
False indigo (*Baptisia* spp.)
Fennel (*Foeniculum vulgare*)
Ferns†
Field scabious (*Knautia macedonia*)
Fleabane (*Erigeron* spp.)
Foamflower (*Tiarella* spp.)†
Foxglove (*Digitalis* spp.)
Foxtail lily (*Eremurus* spp.)†
Gas plant (*Dictamnus albus*)†
German statice (*Goniolimon tataricum*)†
Germander (*Teucrium canadense, T. chamaedrys*)†
Giant hyssop (*Agastache* spp.)†
Globe thistle (*Echinops* spp.)†
Goatsbeard (*Aruncus* spp.)
Gold bleeding heart (*Corydalis lutea*)†
Golden groundsel (*Ligularia* spp.)†
Goldenrod (*Solidago* spp.)
Goldenseal (*Hydrastis canadensis*)
Groundsel (*Senecio* spp.)
Hellebore (*Helleborus* spp.)†
Hens-and-chicks (*Sempervivum tectorum*)†
Hepatica (*Hepatica* spp.)†
Hollyhock (*Alcea rosea*)†

†These plants have proven particularly resistant in IES trials.

Hollyhock mallow (*Malva alcea*)
Horehound (*Marrubium vulgare*)†
Horned poppy (*Glaucium flavum*)
Ice plant (*Delosperma cooperi*)
Inula (*Inula* spp.)†
Iris (*Iris germanica, I. cristata, I. sibirica, I. ensata, I. pseudocorus, I. tectorum*)
Irish moss (*Sagina subulata*)
Ironweed (*Vernonia noveboracensis*)†
Jack-in-the-pulpit (*Arisaema* spp.)†
Jacob's ladder (*Polemonium caeruleum*)
Jerusalem sage (*Phlomis* spp.)†
Joe-pye weed (*Eupatorium* spp.)†
Jupiter's beard (*Centranthus ruber*)†
Knapweed (*Centaurea* spp.)†
Labrador violet (*Viola labradorica*)†
Ladybell (*Adenophora liliifolia*)†
Lamb's ear (*Stachys byzantina, S. officinalis*)†
Lavender cotton (*Santolina chamaecyparissus*)†
Leadwort (*Ceratostigma plumbaginoides*)
Lemon balm (*Melissa officinalis*)†
Leopard's bane (*Doronicum* spp.)
Lily-of-the-valley (*Convallaria majalis*)†
Lilyturf (*Liriope spicata*)†
Lobelia (*Lobelia* spp.)
Lungwort (*Pulmonaria* spp.)†
Lupine (*Lupinus* spp. and hybrids)
Madwort (*Alyssum* spp.)†
Maltese cross (*Lychnis chalcedonica*)
Marguerite (*Anthemis tinctoria*)
Masterwort (*Astrantia major*)
Mayapple (*Podophyllum peltatum*)†
Meadow rue (*Thalictrum* spp.)
Milkweed (*Asclepias* spp.)†
Miniature hollyhock (*Sidalcea malviflora*)
Mint (*Mentha* spp.)†
Mondo grass (*Ophiopogon japonicus*)†
Monkshood (*Aconitum* spp.)†
Montbretia (*Crocosmia* x *crocosmiiflora*)†
Mountain mint (*Pycnanthemum* spp.)†
Mugwort (*Artemisia* spp.)†
Mullein (*Verbascum* spp.)
New Zealand burrs (*Acaena* spp.)
Obedient plant (*Physostegia virginiana*)
Oregano (*Origanum vulgare*)
Oriental poppy (*Papaver orientale*)
Ornamental rhubarb (*Rheum* spp.)†
Ox-eye daisy (*Telekia speciosa*)†
Pachysandra (*P. procumbens, P. terminalis*)
Partridgeberry (*Mitchella repens*)†
Peony (*Paeonia* spp.)†
Perennial blue flax (*Linum perenne*)†
Periwinkle (*Vinca minor*)
Phlox (*P. divaricata, P. stolonifera, P. subulata*)†
Pincushion flower (*Scabiosa caucasica*)†
Pinks (*Dianthus* spp.)†
Prairie coneflower (*Ratibida* spp.)
Prickly pear (*Opuntia humifusa*)†
Primrose (*Primula* spp.)
Purple coneflower (*Echinacea purpurea*)

Pussytoes (*Antennaria* spp.)†
Queen-of-the-prairie (*Filipendula* spp.)
Red-hot poker (*Kniphofia uvaria* and hybrids)
Rockcress (*Arabis* spp.)†
Rodgerflower (*Rodgersia* spp.)†
Rose mallow (*Hibiscus* spp.)
Rosinweed (*Silphium* spp.)†
Rue† (*Ruta graveolens*)†
Russian sage (*Perovskia atriplicifolia*)†
Sage (*Salvia* spp.)†
Sandwort (*Arenaria montana*)
Sea holly (*Eryngium* spp.)†
Sea thrift (*Armeria maritima*)†
Shooting star (*Dodecatheon meadia*)†
Siberian bugloss (*Brunnera macrophylla*)
Siberian wallflower (*Erysimum asperum*)†
Skullcap (*Scutellaria incana*)
Sneezeweed (*Helenium autumnale*)
Snow-in-summer (*Cerastium tomentosum*)†
Soapwort (*Saponaria* spp.)
Speedwell (*Veronica austriaca, V. spicata*)†
Spiderwort (*Tradescantia virginiana*)†
Spurge (*Euphorbia* spp.)†
Statice (*Limonium latifolium*)†
Stonecrop (*Sedum kamtschaticum, S. spurium*)
Sunflower (*Helianthus* spp.)
Sweet cicely (*Myrrhis odorata*)†
Sweet woodruff (*Galium odoratum*)†
Thyme (*Thymus* spp.)†
Tickseed (*Coreopsis* spp.)†
Toadflax (*Linaria* spp.)
Trout lily (*Erythronium americanum*)†
Turtlehead (*Chelone* spp.)
Valerian (*Valeriana* spp.)
Vervain (*Verbena* spp.)
Virginia bluebell (*Mertensia virginica*)†
Wake-robin (*Trillium* spp.)
White gaura (*Gaura lindheimeri*)
Wild ginger (*Asarum* spp.)
Wintergreen (*Gaultheria procumbens*)†
Yarrow (*Achillea* spp.)†
Yellow archangel (*Lamiastrum galeobdolon*)†
Yellow waxbell (*Kirengeshoma palmata*)
Yucca (*Yucca filamentosa*)

BULBS/CORMS

Arum (*Arum italicum*)
Colchicum (*Colchicum autumnale*)
Daffodil (*Narcissus* spp.)
Dalmatian crocus (*Crocus tomassinianus*)
Fritillary (*Fritillaria* spp.)
Ipheion (*Ipheion uniflorum*)
Ornamental onion (*Allium* spp.)
Quamash (*Camassia* spp.)
Snowdrop (*Galanthus nivalis*)
Snowflake (*Leucojum aestivum*)
Squill (*Scilla* spp.)
Winter aconite (*Eranthis hyemalis*)

†These plants have proven particularly resistant in IES trials.

The list above was provided by the Institute of Ecosystem Studies, Box AB, Millbrook, NY 12545; www.ecostudies.org; 845-677-5343. Reprinted with permission.

Southwest

This list was compiled from the observations of gardeners, landscapers, and nursery personnel in north-central Arizona, and is distributed by the University of Arizona.

TREES

Ash (*Fraxinus* spp.)
Cedar (*Cedrus* spp.)
Cypress (*Cupressus* spp.)
Douglas fir (*Pseudotsuga menziesii*)
Fir (*Abies* spp.)
Hackberry (*Celtis* spp.)
Hawthorn (*Crataegus* spp.)
Japanese maple (*Acer palmatum*)
agnolia (*Magnolia* spp.)
aidenhair tree (*Ginkgo biloba*)
ak (*Quercus* spp.)
ne (*Pinus* spp.)
dbud (*Cercis* spp.)
uce (*Picea* spp.)
as mountain laurel (*Sophora* *cundiflora*)
apple (*Acer circinatum*)

UNDCOVERS AND VINES

(*Ajuga* spp.)
f plumbago (*Ceratostigma* *mbaginoides*)
ese spurge (*Pachysandra* *inalis*)
nkle (*Vinca* spp.)
creeper (*Parthenocissus*

S

(many species)
(*Buxus* spp.)

Brittlebush (*Encelia farinosa*)
Buckwheat (*Eriogonum* spp.)
Butterfly bush (*Buddleia* spp.)
Chuparosa (*Justicia californica*)
Cinquefoil (*Potentilla* spp.)
Cotoneaster (*Cotoneaster* spp.)
Currant, gooseberry (*Ribes* spp.)
Dalea (*Dalea* spp.)
Daphne (*Daphne* spp.)
Fairy duster (*Calliandra* spp.)
Firethorn (*Pyracantha* spp.)
Flowering quince (*Chaenomeles* spp.)
Glossy abelia (*Abelia grandiflora*)
Holly (*Ilex* spp.)
Jojoba (*Simmondsia chinensis*)
Juniper (*Juniperus* spp.)
Kerria (*Kerria japonica*)
Lantana (*Lantana* spp.)
Lavender (*Lavandula* spp.)
Leucophyllum (*Leucophyllum* spp.)
Lilac (*Syringa* spp.)
Littleleaf cordia (*Cordia parvifolia* spp.)
Manzanita (*Arctostaphylos* spp.)
Oregon grape (*Mahonia* spp.)
Rosemary (*Rosmarinus officinalis*)
Sage (*Salvia* spp.)
Sumac (*Rhus* spp.)
Turpentine bush (*Ericarmeria laricifolia*)
Viburnum (*Viburnum* spp.)

Lower South

The following list of plants was compiled by Forrest W. Appleton, a retired nursery professional in Bexar County, Texas (USDA Zone 8), and is based on his observations and trials.

SHRUBS

Agarita (*Mahonia trifoliolata*)
Box-leaf euonymus (*Euonymus japonica* 'Microphyllus')
Bush germander (*Teucrium fruticans*)
Ceniza/Texas sage (*Leucophyllum* spp.)
Esperanza (*Tecoma stans*)
Evergreen sumac (*Rhus virens*)
Firebush (*Hamelia patens*)
Goldcup (*Hypericum* spp.)
Gray cotoneaster (*Cotoneaster glaucophylla*)
Japanese boxwood (*Buxus microphylla* var. *japonica*)
Oleander (*Nerium oleander*)
Pineapple guava (*Feijoa sellowiana*)
Pomegranate (*Punica granatum*)
Primrose jasmine (*Jasminum mesnyi*)
Reeve's spirea (*Spiraea cantoniensis*)
Rosemary (*Rosmarinus officinalis*)
Santolina (*S. chamaecyparissus* and *S. rosmarinifolia*)
Sotol (*Dasylirion* spp.)
Texas mountain laurel (*Sophora secundiflora*)
Yaupon holly (*Ilex vomitoria*)
Yew podocarpus (*Podocarpus macrophyllus*)
Yucca (*Yucca* spp.)

PERENNIALS AND ANNUALS

Amaryllis (*Hippeastrum* x *johnsonii*)
Angel's trumpet (*Datura* spp.)
Artemisia (*Artemisia ludoviciana*)
Autumn sage (*Salvia greggii*)
Bearded iris (*Iris* spp.)
Blue plumbago (*Plumbago auriculata*)
Candytuft (*Iberis sempervirens*)
Copper canyon daisy (*Tagetes lemonii*)
Dusty miller (*Senecio cineraria*)
Eupatorium (*Eupatorium coelestinum*)
Garlic chives (*Allium tuberosum*)
Gold-moss sedum (*Sedum acre*)
Hummingbird bush (*Anisacanthus wrightii*)
Jerusalem sage (*Phlomis fruticosa*)
Lantana (*Lantana* spp.)
Larkspur (*Consolida ajacis*)
Madagascar periwinkle (*Catharanthus roseus*)
Mallow hibiscus (*Hibiscus moscheutos*)
Marguerite (*Chrysanthemum frutescens*)
Mealy-cup sage (*Salvia farinacea*)
Mexican bush sage (*Salvia leucantha*)
Mexican hat (*Ratibida columnifera*)
Mexican honeysuckle (*Justicia spicigera*)

Pacific Northwest

Russell Link is an urban wildlife biologist based in Mill Creek, Washington (USDA Zone 8). The list below is based on his observation and study of the deer population in Washington State.

DECIDUOUS TREES

Birch (*Betula* spp.)
Fig (*Ficus carica*)
Little-leaf linden (*Tilia cordata*)
Oregon ash (*Fraxinus latifolia*)
Sumac (*Rhus* spp.)
Willow (*Salix* spp.)

EVERGREEN TREES

Douglas fir (*Pseudotsuga menziesii*)
False cypress (*Chamaecyparis* spp.)
Fir (*Abies* spp.)
Hemlock (*Tsuga* spp.)
Juniper (*Juniperus* spp.)
Oregon myrtle (*Umbellularia californica*)
Pine (*Pinus* spp.)
Spruce (*Picea* spp.)
Tan oak (*Lithocarpus densiflorus*)

DECIDUOUS SHRUBS

Chokecherry (*Prunus virginiana*)
Elderberry (*Sambucus* spp.)
Golden currant (*Ribes aureum*)
Hazelnut (*Corylus* spp.)
Lilac (*Syringa* spp.)
Potentilla (*Potentilla fruticosa*)
Red-flowered currant (*Ribes sanguineum*)
Red-twig dogwood (Cornus sericea)
Snowberry (*Symphoricarpos* spp.)
Spirea (*Spiraea* spp.)
Wild gooseberry (*Ribes* spp.)
Wild rose (*Rosa* spp.)
Winter jasmine (*Jasminum nudiflorum*)

EVERGREEN SHRUBS

Coffeeberry (*Rhamnus californica*)
Evergreen huckleberry (*Vaccinium ovatum*)
Juniper (*Juniperus* spp.)
Manzanita (*Arctostaphylos* spp.)
Mexican orange (*Choisya* spp.)
Mountain laurel (*Kalmia latifolia*)
Mugho pine (*Pinus mugo*)
Oregon boxwood (*Pachystima myrsinites*)
Oregon grape (*Mahonia aquifolium*)
Rabbitbrush (*Chrysothamnus* spp.)
Rhododendron (*Rhododendron* spp.)
Sagebrush (*Artemisia tridentata*)
Silk-tassel bush (*Garrya elliptica*)
Silverberry (*Elaeagnus commutatus*)
Wax myrtle (*Myrica californica*)

GROUNDCOVERS AND LOW SHRUBS

Bunchberry (*Cornus canadensis*)
Cotoneaster (*Cotoneaster* spp.)
Dwarf coyote brush (*Baccharis pilularis*)
Heather (*Erica* spp.)
Juniper (*Juniperus* spp.)
Kinnikinnik (*Arctostaphylos uva-ursi*)
Lithodora (*Lithodora diffusa*)

Oxalis (*Oxalis oregona*)
Salal (*Gaultheria shallon*)
Sunrose (*Helianthemum* spp.)
Wild strawberry (*Fragaria chiloensis*)
Wintergreen (*Gaultheria procumbens*)

PERENNIALS

Baby's breath (*Gypsophila paniculata*)
Bee balm (*Monarda didyma*)
Black-eyed Susan (*Rudbeckia* spp.)
Blanketflower (*Gaillardia aristata*)
Bleeding heart (*Dicentra* spp.)
Blue-eyed grass (*Sisyrinchium* spp.)
California fuchsia (*Zauschneria* spp.)
Catmint (*Nepeta* spp.)
Coneflower (*Echinacea purpurea*)
Coreopsis (*Coreopsis* spp.)
Daisy (*Chrysanthemum maximum*)
Gayfeather (*Liatris spicata*)
Globe thistle (*Echinops exaltatus*)
Hellebore (*Helleborus* spp.)
Iris (*Iris* spp.)
Lobelia (*Lobelia cardinalis*)
Lungwort (*Pulmonaria* spp.)
Lupine (*Lupinus* spp.)
Poppy (*Papaver* spp.)
Red-hot poker (*Kniphofia* spp.)
Rockcress (*Arabis* spp.)
Russian sage (*Perovskia atriplicifolia*)
Sea holly (*Eryngium amethystinum*)
Sea thrift (*Armeria maritima*)
Sedum (*Sedum spectabile*)
Snow-in-summer (*Cerastium tomentosum*)
Solomon's seal (*Polygonatum* spp.)
Wallflower (*Erysimum* spp.)
Wild buckwheat (*Eriogonum* spp.)
Yarrow (*Achillea* spp.)

ANNUALS

Ageratum (*Ageratum houstonianum*)
Bachelor buttons (*Centaurea cyanus*)
Calendula (*Calendula officinalis*)
California poppy (*Eschscholzia californica*)
Clarkia (*Clarkia* spp.)
Cosmos (*Cosmos bipinnatus*)
Larkspur (*Consolida ambigua*)
Sunflower (*Helianthus annuus*)
Sweet alyssum (*Lobularia maritima*)
Zinnia (*Zinnia* spp.)

BULBS, CORMS, AND TUBERS

Corn lily (*Ixia* spp.)
Crocosmia (*Crocosmia* x *crocosmiiflora*)
Crocus (*Crocus* spp.)
Fritillary (*Fritillaria* spp.)
Trillium (*Trillium* spp.)

HERBS

Chive (*Allium schoenoprasum*)
Garlic chive (*Allium tuberosum*)
Hyssop (*Hyssopus officinalis*)
Lavender (*Lavandula* spp.)
Mint (*Mentha* spp.)
Oregano (*Origanum vulgare*)
Rosemary (*Rosmarinus officinalis*)
Rue (*Ruta graveolens*)
Santolina (*Santolina* spp.)
Sweet marjoram (*Origanum majorana*)
Thyme (*Thymus* spp.)

The list above is excerpted from Russell Link's book *Landscaping for Wildlife in the Pacific Northwest* (University of Washington Press, 2004). Reprinted with permission.

Mexican marigold (*Tagetes lucida*)
Mexican oregano (*Poliomintha maderensis*)
Ox-eye daisy (*Chrysanthemum leucanthemum*)
Philodendron (*Philodendron selloum*)
Prickly-pear cactus (*Opuntia* spp.)
Salvia (*Salvia* 'Indigo Spires')
Soapwort (*Saponaria officinalis*)
Texas betony (*Stachys coccinea*)
Texas swamp mallow (*Pavonia lasiopetala*)
Thyme (*Thymus* spp.)
Wormwood (*Artemisia absinthum*)
Yarrow (*Achillea millefolium*)
Zinnia (*Zinnia* spp.)

ORNAMENTAL GRASSES

Gulf muhley (*Muhlenbergia capillaris*)
Lindheimer's muhley (*Muhlenbergia lindheimeri*)
Maiden grass (*Miscanthus sinensis*)
Purple fountain grass (*Pennisetum setaceum*)
Sea oats (*Chasmanthium latifolium*)

The list above was originally published by Texas A&M, at its extension Web site, http://aggie-horticulture.tamu.edu/plantanswers/publications/deerbest.html, and is reprinted with permission.

PERENNIALS, BULBS, AND ANNUALS

Agave (*Agave* spp.)
Artemisia (*Artemisia* spp.)
Aster (*Aster* spp.)
Beardtongue (*Penstemon* spp.)
Bee balm (*Monarda* spp.)
Begonia (*Begonia* spp.)
Bellflower (*Campanula* spp.)
Blackfoot daisy (*Melampodium leucanthum*)
Blanketflower (*Gaillardia grandiflora*)
Bleeding heart (*Dicentra* spp.)
Blue fescue (*Festuca ovina* 'Glauca')
California fuchsia (*Zauschneria californica*)
California poppy (*Eschscholzia californica*)
Candytuft (*Iberis* spp.)
Catnip (*Nepeta* spp.)
Centaurea (*Centaurea* spp.)
Columbine (*Aquilegia* spp.)
Coreopsis (*Coreopsis* spp.)
Cranesbill (*Geranium* spp.)
Crocus (*Crocus* spp.)
Crown-pink (*Lychnis coronaria*)
Daffodil (*Narcissus* hybrids)
Dahlia (*Dahlia* hybrids)
Dead nettle (*Lamium maculatum*)
Euphorbia (*Euphorbia* spp.)
False spirea (*Astilbe* spp.)
Feather grass (*Stipa* spp.)
Ferns (many species)
Fleabane (*Erigeron* spp.)
Forget-me-not (*Myosotis scorpioides*)
Gloriosa daisy (*Rudbeckia hirta*)
Impatiens (*Impatiens* spp.)
Iris (*Iris* spp.)
Lamb's ears (*Stachys byzantina*)
Lavender cotton (*Santolina* spp.)
Lupine (*Lupinus* spp.)
Moss pink (*Phlox subulata*)
Naked lady (*Amaryllis belladonna*)
Oregano (*Origanum* spp.)
Oriental poppy (*Papaver* spp.)
Pincushion flower (*Scabiosa* spp.)
Red-hot poker (*Kniphofia uvaria*)
Saxifrage (*Saxifraga* spp.)
Sea thrift (*Armeria* spp.)
Snow-in-summer (*Cerastium tomentosum*)
Speedwell (*Veronica* spp.)
Squill (*Scilla* spp.)
Strawflower (*Helichrysum bracteatum*)
Swan river daisy (*Brachycome iberidifolia*)
Sweet violet (*Viola odorata*)
Thyme (*Thymus* spp.)
Verbena (*Verbena* spp.)
Yarrow (*Achillea* spp.)

The list above was issued July 2001 from the University of Arizona, College of Agricultu and Life Sciences, Cooperative Extension, P.O. Box 210036, Tucson, AZ 85721-0036. Reprinted with permission.

Roses That Rise to the Occasion

There is hope for the die-hard rose enthusiast in deer country. When roses become deer food, it's *always* new growth that suffers the most deer damage. New growth is the most tender and supposedly tastes the best. Hence, that whole "browsing" thing . . . just take a little off the top. You can limit the damage by covering the plant with the same netting that is used to protect fruit, such as blueberries, from birds. Although every rose is subject to deer predation at some level, those listed below have repeatedly sustained the least damage.

SPECIES ROSES

Rosa villosa. Mauve-pink, semidouble, extremely fragrant blossoms. Tall, thorny shrub. Commonly referred to as the "apple rose," because of its large, applelike fruit. A layer of hairs around the seeds can cause irritation to the mouth and digestive tract if ingested. Likes well-drained soils, but tolerates clay. Grows well with alliums, parsley, and lupine, but poorly near boxwood. Zones 5–9

Rosa sericea pteracantha. Tiny pink blossoms. Canes bear enormous, deep red thorns. Grows best in the Pacific Northwest, mid-Atlantic region, and temperate parts of the Midwest, the Rockies, and the Northeast. Good as a barrier hedge. Zones 5–8

Rosa rugosa. Mauve-pink blossoms with heavy scent. Thick, dense foliage. Covered with thorns. A tough, disease-resistant species. Tolerates shade and salt. Many cultivars are available, including 'Alba' (white) and 'Rubra' (red). Zones 2–9

Rosa soulieana. White blossoms. Very thorny; exceptionally tall. A unique and beautiful rose with a strong growth ethic. Canes can be trained as climbers or as a pillar, or allowed to arch naturally into a large (10-by-10-foot) shrub. Foliage is disease resistant. Canes bear yellow thorns; bloom is generous through midsummer and is followed by orange-red hips. Zone 7

Rosa pimpinellifolia. White to yellow blossoms. Very thorny; fernlike leaves. Grows best in cooler regions such as the Northeast, the Rockies, and the Pacific Northwest. Good for hedges and is an excellent garden companion. Plant with deer-resistant herbs like artemisia, nepeta, sage, lavender, and rosemary. Zones 3–8

RUGOSA HYBRIDS

Rugosa hybrids as a group are tough plants that survive a variety of conditions, from seaside or city environments to nibbling by curious deer. The flowers often have a strong clovelike scent, that, although pleasant to us, apparently doesn't appeal to deer. Canes are usually very thorny, leaves are leathery and tough, making for unappetizing deer fodder.

'Agnes'. Light yellow; fragrant blossoms that continue through mid-spring. Resistant to black spot and mildew, but susceptible to rust. Zones 6–9

'Belle Poitevine'. Pink blossoms with intense fragrance. Blooms continuously in late spring through early summer. Zones 4–8

'Blanc Double de Coubert'. Fragrant white blossoms. Disease resistant, shade-tolerant. Flowers on thorny stems; large hips for fall color and winter interest. Zones 3–9

'Delicata'. Mauve-pink blossoms. Dates to 1898, blooms in late spring/early summer. Disease resistant and well-armed with thorny branches. Blooms on new wood, so early pruning promotes new flowering growth. Zones 3–9

'Hansa'. Red-violet blossoms. Good as a hedge. Flowers are very fragrant with a spicey, clovelike aroma. Disease resistant and prefers no pruning. Zones 3–8

'Scabrosa'. Red, fragrant blooms from late spring through early summer on very thorny stems. Disease resistant. Zones 3–8

'Therese Bugnet'. Red-violet blooms are double, though only slightly fragrant, and continue from late spring through early summer. Shade tolerant and resistant to black spot. Excellent for northern gardens. Zones 3–9

SHRUB ROSES

'Baronne Prevost'. Very fragrant pink blossoms start in early summer and keep coming until fall. Susceptible to black spot. Grows quickly. Zones 4–9

'Conrad Ferdinand Meyer'. Pink, intensely fragrant blooms. Very vigorous. Dates from 1899. Grows up to 10 feet tall. Tea-rose-type blossoms repeat from late spring through early summer. Zones 5–10

'Fisherman's Friend'. Crimson blooms. An English rose that grows to about 4 or 5 feet tall and just as wide with huge blossoms that continue in waves throughout the summer. Zones 5–10

'Harison's Yellow'. Deep yellow blooms. Introduced in 1830, 'Harison's Yellow' came west with the pioneers. Forms a dense shrub or can be trained to climb. Blooms in late spring through early summer. Shade tolerant. Zones 3–9

'Penelope'. Coral-pink blooms with outstanding fragrance. Upright; tall. Bushy; excellent hedge. A hybrid musk rose that grows to 8 feet tall. Can be trained to climb a support. Blooms from late spring to early summer. Color is best during the coolest part of the bloom season. Dislikes chemical sprays and pruning. Zones 5–10

'Robusta'. Red blooms. Thorny; vigorous growth; very tall. Another old-fashioned favorite, dating back to 1877. This Bourbon rose produces very fragrant flowers in late spring to early summer. Zones 5–10

OLD GARDEN ROSES

'Alfred de Dalmas'. Light blush-pink blooms. Very bristly, thorny; rough leaves. A form of moss rose, so called because of the fuzzy, sticky sepals surrounding the flower buds, which ruin the whole appeal of roses for deer. Introduced in 1855. Grows 3 to 4 feet tall. Zones 4–9

Common Moss Rose. Pink, fragrant blooms in midsummer. Very thorny. Fuzzy calyxes. Blooms on old wood, so prune after flowering. Many varieties dating from the early to mid-1800s. Zones 5–9

'Crested Moss'. Very fragrant, pink blooms open from "mossy" buds in midsummer. Very thorny. Rough leaves. Prone to weak stems. Zones 4–9

'Henri Martin'. Dark red, fragrant blooms. Thorny, bushy, and vigorous. Another moss variety, from 1852. Grows to 8 feet tall and 6 feet wide. Blooms from late spring to early summer. Disease resistant. Zone 4-9

'General Kleber'. Very fragrant medium pink blooms from late spring to early summer. Thorny; bushy. Rough leaves. A compact moss rose dating from 1856. Grows to about 4 feet by 4 feet. Blooms are so full they are "quartered," separating into sections. Zones 4-9

'Konigin von Danemark'. Very fragrant pink blossoms from late spring to early summer. Very thorny; rough leaves. Introduced in 1826 and still a favorite. An Alba class rose that is excessively bristly. Disease resistant. Zones 4-8

'Louis Gimard'. Mauve-pink blooms in late spring to early summer. Very thorny; pine-scented, mossy buds. A moss rose introduced in 1877. Bushy habit, growing to 6 feet tall. Disease resistant. Zones 4-9

'Madame de la Roche-Lambert'. Pink, globe-shaped blooms from late spring through early summer. Very thorny, arching canes with rough leaves. Dates from 1851. Resistant to black spot and rust, but susceptible to mildew. Zones 4-9

'Maiden's Blush'. Light blush-pink blossoms. Very fragrant; bristly; thorny; tall. A hybrid Alba rose, with an upright habit. Good for hedging. Shade tolerant. Introduced in England in 1797. An oldie *and* a goodie! Zones 4-9

'Striped Moss'. Red-and-white-striped flowers. Very thorny; rough leaves. Interesting striped blooms make this moss stand out in the garden, and nasty thorns, ragged foliage and mossy buds keep deer away. Zone 4-9

'William Lobb'. Deep mauve, semidouble to double, very fragrant blooms in midsummer. Bristly, thorny; mossy; heavily scented. Another moss rose variety, growing to 8 feet or more. Can be trained up a support. Zones 4-9

'York & Lancaster'. Pink-and-white-striped petals on fragrant flowers in late spring and early summer. Very bristly, thorny; rough leaves. Tolerates partial shade. Zones 3-9

Testing for Palatability

It figures. You scoured the lists, and the plants you're concerned about aren't mentioned. Don't fret; over time a lot of your knowledge about deer will come, not from lists, but from your own observations and experimentation. For example, to determine if a new plant will be considered bait or bane by neighborhood deer, place a sample in a pot and leave it where deer are known to feed. (Actually, it makes sense to start by leaving out just the empty container for a few days, to get the deer accustomed to something new in their territory.) For small plants, this could be an entire potted specimen; for larger plants, try a cutting or branch. If the deer leave it alone, it should be safe to plant. If they devour it, try something else.

When considering new plants for the garden, compare them with related plants. Often this will provide a clue as to whether they're likely to wind up as deer fodder. Pay particular attention to wild plants in your area that deer leave alone. For instance, deer don't eat mayweed (*Anthemis cotula*), and will also shun such relatives as golden marguerite (*Anthemis tinctoria*). Buttercups (*Ranunculus* spp.) are toxic to deer and therefore avoided, as are closely related plants like rocket larkspur (*Consolida ambigua*), every part of which is poisonous, and peonies. Other examples are the noxious weed toadflax (*Linaria* spp.), which deer ignore, and its gloriously blooming garden relative, the snapdragon. Foxglove (*Digitalis purpurea*) is a toxic plant avoided by deer even in its cultivated forms. Deadly datura and its cherished southern garden relative angel's trumpet (*Datura* spp. and *Brugmansia* spp.), as well as the perennial bindweed (*Calystegia sepium*), are avoided for their unpalatable taste. On the other hand, most members of the rose family (Rosaceae) are like deer candy; related plants, such as apples, cherries, and raspberries, all suffer the same fate when deer venture by.

Can Poisonous Be Practical?

Be practical with poisonous plants. People resort to growing poisonous plants in an effort to grow *something* that deer won't consume. But there are two calculated risks here. First, being ruminants, deer can tolerate a lot more toxicity than people, pets, and most livestock can. Even a plant that tastes horrible and makes them sick will be consumed if deer are hungry enough. Deer will eat anything — including the paper deer-resistant plant lists are printed on — when starving. Second, anytime you cultivate toxic plants such as monkshood (*Aconitum nepallus*), castor bean (*Ricinus communis*), yew (*Taxus* spp.), or nightshade (*Solanum* spp.), be aware of the dangers they present to people and pets. These and other toxic plants can cause accidental poisoning if any part of the plant is eaten. Carefully gauge whether you should grow poisonous plants simply to discourage deer, as the dangers these plants pose may not be worth their benefits.

Tricks of the Trade

When you design a garden to avoid deer damage, the most important considerations — the purpose of your garden and the overall look you desire — don't change. But you can accomplish most goals in more than one way. Many of your favorite plants that deer greedily scarf down have relatively deerproof counterparts that can fulfill the same function. Do you long for spring-flowering bulbs? Then ditch the tulips and deploy daffodils. There are at least 60 varieties of daffodils in 13 divisions with a range of colors from pearly white to brilliant yellows and peaches, and an array of attractive flower forms. They grow in a variety of sizes with a range of bloom times to fill any garden gaps left by tulips. Look to alliums to provide a palette of blues, and many irises to color coordinate any deer-resistant garden design. Need a shady spot in the summer yard? Fig, birch, horse chestnut, and mature common lilac trees will fill the bill, not a deer's stomach. Need a hedge for privacy? Forget about such deer delicacies as hemlock and yew and concentrate on boxwood or rugosa roses.

Once you compile a list of suitable substitutes, factor in any plants you really can't live without. Careful design (and perhaps some of the training techniques listed in the next chapter) should allow you to enjoy deer-tempting favorites with little or no damage. The purpose of deer-o-scape design is to make it appear to the deer that you grow only fuzzy, stinky, yucky-tasting stuff and that they may as well seek supper elsewhere. Remember, being the creatures of habit that they are, once deer determine that your yard is absolutely tasteless, you'll have less trouble with returning foragers.

Do you long for spring-flowering bulbs? Then ditch the tulips and deploy daffodils.

Try these deer-o-scaping tricks:

Substitute the unsavory. As mentioned, for nearly every plant that deer pilfer, there's bound to be at least one substitute that deer detest. Though hybrid tea roses are favored fare, an intense scent, thorny canes, leathery leaves, and a propensity to sprawl into impenetrable barriers combine to make many rugosa hybrids nearly deerproof. Categorize the plants in your garden by function, form, color, and so forth, then shop the lists of plants deer avoid for suitable alternatives.

Create uninviting entryways. Whether through the driveway, an open field, or an alley, it's a good bet that deer enter your property the same way every time. Make sure deer find the entryway to your garden unattractive. Concentrate deer-repelling plants here. This is also a good place to utilize repellents and other deterrents. Conversely, sometimes just moving susceptible plants away from deer's main traffic areas is enough to avoid damage. Deer tend to keep to the beaten path, and moving daylilies to the road less traveled may be all it takes to save them.

Foul the fringes. Keep your "edges" equally unattractive. Line *your* territory with unpalatable and repellent plants and chances are the deer will keep to *their* territory and bypass your no-longer-tempting yard. "What's the use? This place stinks," is what you want the deer to think. How heavily concentrated such plants should be along the edge of your property depends on how heavy the deer pressure is in your area. It could be that your only defense is a wall-to-wall pieris hedge. But a less drastic approach works for most gardeners. A predominately unattractive border is enough to deter most deer, most of the time. With their senses overwhelmed with the scent of yarrow or aromatic herbs, a mouthful of fuzzy-leafed lamb's ear, or a bitter bite of oleander, most deer will follow their nose and their eat-as-they-go habit and keep moving in search of more-appetizing fare.

Create confusing combinations. "I was sure there were daylilies in there somewhere," one deer says to another, "but all I could smell was garlic." The moral: Deer won't eat what they can't find. Surrounding and interplanting susceptible plants with unpalatable or repellent plants makes them much more difficult for deer to find, and — guilt by association — much less attractive if they are detected. A combination of strongly aromatic plants mixed throughout the landscape jams the deer's best radar for finding his favorites.

Deploy deceptive defenses. Camouflage smaller plants with deterrent companions. Larger plants, such as young trees and shrubs, can be defended by surrounding them with unappetizing companions. Surround or interplant early-blooming crocuses or tulips with deer-resistant lily-of-the-valley, squill, or primrose. Underplant young trees with deterrent aromatic mints and plant a screen of astilbe, foxglove, delphinium, spurge, or ornamental grass around them. Just be sure the trees receive enough light for growth. (Some thinning of cover plants may be necessary as the season wears on.)

Provide no view. The portions of your yard that are obscured from sight are much less likely to be invaded than are those with a clear view. Create a deer barrier from a garden border using solid hedges grown from rugosa roses or boxwood, or use trellises swathed in morning glories. Deer won't venture past anything they can't see through or over.

Eliminate unnecessary cover. Tall grass, brushy borders, and areas of your property left to grow wild all encourage deer to bed down. Keep grass and underbrush trimmed and tidy near the garden to discourage loitering.

Be tidy. It's never wise to leave fruit rotting on the vine or tree. Few things entice deer like ripe apples and pears, so be a tidy gardener. Keep fruit picked up (this also reduces the threat of other pests, including yellow jackets) and remove or till under the remains of deer-favored crops, such as corn and peas, as soon as the harvest is finished.

Provide no landing site. Deer won't attempt to leap into your yard if it appears they have no landing site. This depends more on the lay of your land than other deerproofing techniques, but incorporating terraces and multiple levels around the perimeter of your plot may discourage deer from alighting on your lawn.

Adapt self-sustaining garden techniques. Xeriscaping is a gardening style tailored to the needs of gardens under chronic drought conditions. The tenets are simple: Do more with less and conserve water. A self-sustaining garden similarly does more with less and uses the resources where you live to best advantage. No endlessly rolling stretches of green on green, these gardens are designed to look natural, yet defined. Gravel pathways of contrasting-colored rock serve as borders and punctuation lines. Attractive mulches (most of which deer don't eat) conserve water and draw the eye. Plants tolerant of the climate in your neck of the woods grown in carefully planned positions give the appearance of more greenery than there actually is.

Try tasteless lawn ornaments. Just as self-sustaining gardens take the focus off many green growing things and redirects it toward specifically designed points of interest, so too can other gardening styles. You don't necessarily need a full flock of flamingos to replace the visual appeal normally supplied by greenery. Garden seats, sundials, sculptures, and fountains make wonderful focal points in the landscape. Consider removing any deer-luring plants and replacing them with nonedible points of interest.

Garden Designs That Discourage Deer

A deer-o-scaped yard has few limitations. Regardless of what you seek from your garden sanctuary — color, texture, aromas, privacy, functional space or just good old peace and quiet — you can achieve it with a design that also deters deer. It may take a little trial and error, as deer are constantly making adjustments to our ways of doing things, but the yard of your dreams is really just a growing season or two away. Let your preferences be your guide and follow these four steps to create your deer-resistant garden:

Identify and accept the things you can't change, such as temperature extremes, overall soil type, and yearly rainfall. These are your most limiting factors. Even though oleander and angel's trumpet offer near guaranteed deer resistance, they are tender perennials that rarely survive outside of Zones 8 to 10. However, even though Russian sage (*Perovskia atriplicifolia*) will survive Zones 3 to 9, this plant loves dry, alkaline soil and usually can't even get a foothold in soggy Pacific Northwest gardens. Pick plants that will thrive in your conditions. The healthier the plant, the more likely it is to withstand browsing, even if its other natural deer defenses (bad smell, hairy leaves, and so on) fail it.

Define your space. Decide where play, parking, and garden areas belong. This is important, as gravel and lawn are much less alluring and much less susceptible to deer than is a berry patch. If possible, situate parking or play areas (complete with playground equipment and squealing kids) where approaching deer are sure to notice them first. Try to keep the most vulnerable areas of your yard protected from the outside by layers of other uses. Start with a parking area, for instance, then playground, then ground cover, then lawn. The areas that are least susceptible to deer damage should be what they encounter upon approaching your yard.

Look up, look down. Incorporate vertical plants and groundcover for variety. A deer-resistant garden need never be dull or uninspired. In fact, incorporating a range of deer-repellent and/or -resistant plants at all levels will help add interest to your space and will reassure deer that every inch of your yard is a wasteland of bad taste. Prostrate rosemary (*Rosmarinus officinalis* var. *prostratus*) makes an attractive, fragrant, spreading groundcover that complements the whites and violets of old-fashioned climbing morning glory, or eave-clinging, perfumed wisteria. Adding terracing or sunken beds to your yard helps to deter deer, as they dislike climbing in and of steep grades or confined spaces.

Indulge your personal taste. Choose the colors, textures, and scents you prefer. Though you may not be able to cultivate exactly the garden you had envisioned before a deer problem reared its fuzzy-antlered head, you can come surprisingly close to achieving the same effects by choosing color and form over specific plants. Alternative choices are abundant. In fact, the most challenging aspect of designing a garden that won't appeal to deer may just be narrowing down your choices. Peonies mimic old-fashioned roses in form, and poppies carry on blooming after the peonies have faded. Irises and alliums sport the range of colors of tulips, though they bloom later in the season. The variety of color choice, plant type, growth habit, blooming time, and fragrance among deer-resistant or deer-repellent plants is extensive.

Each of these steps will help you progressively narrow down the choices on the deer-resistant plant list. Use the remaining options as the buildings blocks of your deer-o-scaped yard. You can interplant other, nonresistant favorites with relative safety, especially if you wait until the deer become accustomed to your unappetizing choices. Deer know they can't afford to keep coming back to the same garden in search of food if there is nothing good to eat there. Even in times of plenty, their instincts tell them to move on.

FIVE

Deer Deterrents

With the mysteries of what deer eat and why now revealed, we can get down to convincing them they don't want whatever is growing in your yard. There are several ways to accomplish this, and the effectiveness of any one method depends on several factors, including the deer pressure in your area, the surrounding environment, and the established habits of "your" deer, as well as the plants you have in your yard and garden.

If you've applied the deer-o-scaping principles and suggestions discussed in chapter 4, you should be on your way to curbing your deer problem. Unfortunately, even strict adherence to the deer-o-scaping guidelines won't guarantee complete success, and is of only limited value to those of you with well-established plantings that you are loath to uproot. So, what next?

Deer-o-scaping represents benign intervention, a gentle approach that is a good start but most often not a complete solution. To make your garden even more deer resistant requires that

Why Deterrents Do or Don't Work

Just as some teachers (and students) are better than others, some deterrents are more effective at training deer in certain situations than others. Here is a summary of points that largely determine whether deterrents do or don't work:

• A deer's first priority is not to get eaten, and in order to avoid becoming prey, deer have a well-developed predator-avoidance system. Deterrents that capitalize on that system are very effective at keeping them out of the garden.

• Deer in wild country react differently from deer that have adjusted to the presence of humans. Suburban gardeners have to be a tad more resourceful than their country cousins simply because their presence alone is not considered threatening to deer in the developments, as it is to their wild cousins.

• Deer are creatures of habit. Their previous experience with any given food or deterrent dictates their future response to the same or similar circumstances. *Preventing deer damage before it starts is easier than interrupting an established pattern.*

• Deer are, however, adaptable. No matter how effective a deterrent may be when you first employ it, chances are that unless it jumps up and comes after them, the deer will invariably get wise to the fact that it can't eat them. And once they adapt to your garden, they adopt it. That's what makes changing tactics *before* the deer get wise to them a crucial component of employing detterents.

Just as a plant in one garden may survive deer but be eaten in a garden a few miles away, deterrents that work fabulously in one setting may not help at all in another, or even in the same setting at another time. Differences in how deer react to deterrents occur from species to species, region to region, season to season, individual to individual, and for no apparent reason. Whitetails tend to adapt to human activities more quickly than deer whose habitat we have not shared for as many generations, such as blacktails. But all deer are maddeningly adept at adapting to our efforts to keep them at bay.

Five Sensible Approaches

The individual variability of deer and their enviable adaptability would be problem enough, but you also have to contend with their superior physical senses. Given all this, it may seem as though they have the advantage against gardeners merely trying to protect their plants. Well, maybe they do. But remember that on your worst day, you're smarter than deer. You can use your superior reasoning ability to turn a deer's strengths into weaknesses by exploiting those superior senses and gaining the results you want. You *can* succeed.

The deer's five physical senses give us five different ways to assault their feeling of security and send them scampering back to the woods (or at least over to the neighbor's yard). Like all wild animals, deer are neophobic (afraid of anything new). The strange and the unpredicable spell trouble to a prey species. The trick, then, is to not give them a chance to adapt. Thus, the most important advice in any deer deterrent program: Plan on using *several* deterrent tactics, and rotating and alternating them throughout the course of the season, *before* the deer get used to them. That way, mysterious and frightening deterrents become all the more mysterious and frightening.

Assaulting all five senses also gives you some choices as to what you can tolerate in your yard, for some of the things suggested to repel deer are equally effective against gardeners. Always weigh the severity of the deer damage against the inconvenience, unsightliness, and/or cost of the deterrents.

Key Deterrent Strategies

- Use several deterrent tactics.
- Rotate tactics throughout the season.
- Change tactics before deer get used to them.
- Take advantage of all the deer's senses.

Foul Smells

We know that deer have a very sensitive sense of smell and rely heavily on it. Thus, we can be real stinkers (to the deer) when we need to be. There are two strategies to deter deer through scent: jamming their sensors, so to speak, and setting off a red alert. Both approaches call for area repellents — that is, disseminated repellents that carry lingering odors. How big an area is protected depends on the repellent and how you use it.

Aroma deterrents that jam the deer's sensors have such a strong odor that deer in their vicinity have trouble scenting through them. These smells aren't necessarily offensive to humans, just intense. Not being able to scent the wind for danger is an uncomfortable situation for deer, and they can't tolerate it for long.

"Red-alert" deterrents offer a more direct approach and are more effective, when used properly, than masking scents. These are predator scents that, rather than block the whispers of other scents on the wind, scream, "Run for your life!"

Bear in mind that garden conditions may require frequent applications to keep repellent scents fresh and effective. Most need reapplication after a heavy rain, though humid conditions actually enhance odors. Don't forget that deer feed from ground level to as high as six feet (that's six feet above the *snow line* in winter) and that repellents must be applied within that range. Many hold their repellent qualities longer if protected in containers. Some suggestions for homemade containers are included with the specific repellents discussed below, but savvy garden suppliers have designer containers available.

Early-spring applications protect the garden as new growth begins to emerge. Reapplications keep up the level of protection throughout the growing season. And if applied in late fall, before temperatures dip below freezing, repellents protect evergreens and other vulnerable plants from winter browsing.

Finally, remember that deer get used to new odors and may decide, eventually, that any given odor repellent is not much of a threat. Be sure to keep them guessing by changing the location and type of repellent from time to time.

Soap

Scented soaps repel deer. Just who discovered this is not known, but leaving bars of soap about the garden scares away deer. Bar soaps and soap-based commercial repellents made from higher fatty acids are both fairly effective in repelling deer.

To use: Drill a small hole just large enough for a cord to pass through in a wrapped bar of soap (the wrapper keeps the soap from quickly washing away). Then tie with string and hang about the yard, in tree or shrub branches at a height that best protects the plant in question; three to six feet high for most. Alternatively, place unwrapped soap in cheesecloth bags or old nylon stockings and tie the sudsy little sachets around the garden. They will wash away more quickly, but the scent will be more intense and lingering.

Positioning of the soap is important to prevent deer from browsing between the bars. Yes, scientific tests have actually been conducted to determine just how close deer can stand to feed next to soap, and the consensus is about three feet. That comes to

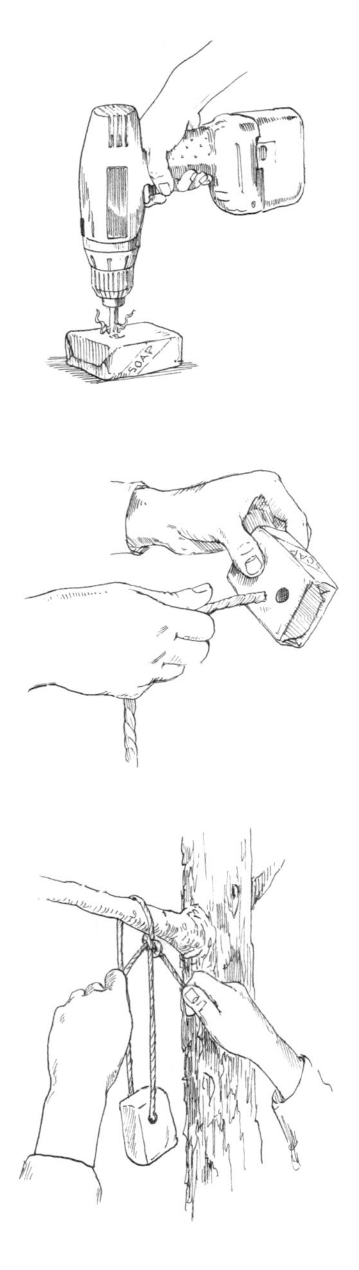

Leaving the wrappers on, hang bars of soap throughout the garden or on individual trees and shrubs.

approximately 450 bars to the acre. For a large garden, that's a lot of soap. Many home gardeners find the best use of soap is hanging it in fruit or ornamental trees whose branches may fall victim to deer damage.

Which brand of soap you choose doesn't appear to be critical. Some people insist that heavily scented deodorant soaps, such as Irish Spring, are most effective. However, avoid soaps containing edible oils as a main ingredient, because at least one report recounted deer actually eating soap made of coconut oil. A 1992 Cornell University study discovered that the repellent component of soaps is tallow (animal-origin) fatty acids. And not surprisingly, it confirmed that soaps containing coconut fatty acid were not effective as repellents.

Pros and cons: Hanging soap in the garden is fairly effective in preventing deer from browsing within three feet of the bar. It's inexpensive (on a per-tree basis), safe, and easy to install, and the bars need not be replaced until they have nearly been washed away by rain or sprinkler water.

One problem, however, is that the same soap fats that repel deer can attract rodents. As soap washes down the branches and trunks of trees and shrubs, it lures rodents to chew on the bark. To prevent rodent damage, combine the use of soaps with a rodent repellent. Hot pepper mixes will repel both the big browsers and the little nibblers.

Related commercial products: Soap-based commercial preparations include Hinder, and DeerBusters. Unlike many such deterrents, Hinder is approved by the United States Department of Agriculture to be applied directly to plants that are to be eaten, such as vegetables and fruit trees. The strong ammonia smell is regarded as almost twice as effective as hanging bars of soap, but Hinder does not weather as well, lasting just two to four weeks. For best results, apply it with an antidesiccant, such as Wilt-Pruf or Vapor Gard.

Hair

Human hair is saturated with our scent and as such proves to be an effective deterrent against wild deer. More-urbanized deer, however, are far too sophisticated to be put off by this, and using human hair as a repellent is generally a wasted effort. If you want to give it a try, pay your local hairdresser a visit with empty bags in hand and offer to sweep up. Hair from predators, when available, works well. No need to shave a lion: Dog groomers sweep up and discard buckets of predator hair.

To use: Stuff hair or fur into cheesecloth bags or sections of old nylons and tie closed. Place the hair about the garden and/or hang from trees and shrubs. As with soap bars, place the hair bags no more than three feet apart. Hang bags about three feet off the ground. On large trees, hang them from perimeter branches.

Pros and cons: The pros are that hair is free, easy to come by, and usually works to deter deer. The cons are that some deer will ignore the smell, and that even if the hair scent does work, it must be replaced at least once a month, as it gradually loses its odor.

Repellent Plants

Interplanting repellent plants with more-vulnerable species is a valuable technique in the art of deer-o-scaping. Heavily scented herbs, such as artemisia, lavender, Russian sage, tansy, and yarrow, as well as culinary herbs, including thyme, tarragon, oregano, dill, and chives, often prove intolerable to deer. Oddly, plants toxic to deer, such as foxgloves, don't appear to have repellent properties, as deer frequently feed among them.

To use: Refer to chapter 4.

Pros and cons: Although strategic plantings call for a good deal of thoughtful planning, they can succeed in discouraging deer traffic in areas of low-to-moderate deer pressure. Specific herbs and deterrent plants are readily available through mail-order outlets, if not closer to home.

Spray It On!

For most effective use of spray-on deer repellents, consider the following tips:

- Add an antidesiccant/surfactant product such as Wilt-Pruf or Vapor Gard to home brews to extend the life of the spray.
- Reapply often in the spring and early summer to protect new growth. New growth is most tender and sought by deer, and left unprotected will be the first to go.
- Reapply as often as recommended to ensure coverage and efficacy.
- Keep an eye on how deer respond to repellents; rotate, change, combine, and/or modify *before* they adapt to repellent odors or tastes.

Garlic and Rotten Eggs

The sulfur compounds contained in garlic and fermented egg solids are highly repellent to deer and other garden pests. Deterrents made from these ingredients are among the most effective scent repellents available. Egg solids are probably the most popular and safest of the ingredients in deer repellents. The Environmental Protection Agency lists them as minimum risk. You can use them on all plant types and they last from one to three months. Commercial products include Deer Away and Deer Off.

To use: Concoct your own homemade version by processing one egg per cup of water in a blender or with a hand mixer. Leave the slurry to ferment for several days before applying to plants. Bits of egg may remain suspended in the mixture, which can clog sprayer nozzles, so the best way to apply the solution is to pour or sprinkle it over the foliage of vulnerable plants. Add a clove of garlic to the mix and a dash of chili pepper or Tabasco sauce for extra punch, to double its effectiveness.

Another option is to stuff old socks or stockings with crushed or bruised cloves of garlic and hang them in shrubs or trees. Garlic is also used as an insect and rabbit repellent, and is listed as minimum risk by the EPA. It easily washes off, making it good for protecting vegetables; however, it doesn't last very long. Garlic socks are most effective during warmer months.

Pros and cons: Fermented egg solids and strong garlic scents rank among the most repellent odors out there, short of hot lion's breath, and are most effective in areas of low to moderate deer density. These aromas are so pungent that you may not be able to stand the smell of the mixture as you prepare or apply it. Fortunately, after the egg mixture dries on the plants and surrounding soil, people find the aroma much less noticeable, but the deer's superior sense of smell will still detect it.

Related commercial products: The proliferation of commercial deer repellents based on the sulfur compounds found in garlic and eggs offers substantial evidence that these scents effectively deter deer. Examples include Deer Away (also sold as Big Game Repellent, or B.G.R.) and Deer Off. Deer Away is registered for use on fruit trees (before flowering, when dormant, or during nonbearing times), nurseries, ornamentals, and conifer seedlings, but is not approved for food crops. It maintains its effectiveness for up to a month with a single application, regardless of rainfall. Deer Off also clings after a rain. Also, wettable sulfur, sold as a disease-fighting agent or soil amendment, gives off a similar odor. Follow label instructions.

Eau de Rotten Eggs

- 1 - 4 garlic cloves
- 2 cups water
- 2 eggs
- 2 tablespoons Tabasco sauce (optional)

At high speed in the blender, purée the garlic in the water. Add the eggs and Tabasco and process mixture thoroughly. Allow the mixture to rest, covered, for several days prior to application.

The big advantage to commercial products is that you don't have to mix them up yourself. Also, they can be sprayed, making them easier to use and ensuring more-even application to foliage than sprinkling the homemade brew. Uniform application is less of a concern, however, with scent repellents than with taste, or contact, deterrents.

Chicken Feathers

Which came first, deer's aversion to eggs or to chicken feathers? Add this by-product of the poultry industry to the list of things deer find repugnant.

Feather meal is a slow-release, high-nitrogen fertilizer (12-0-0), often used as part of organic fertilizer mixes. It differs from many other nitrogen fertilizers in that feather meal contains nitrogen protein rather than nitrates or ammonia. It is also considered very effective at repelling deer. A 2000 report released by Michigan State University found that feather meal was 90 to 100 percent effective in deterring deer. But true to deer form, a Cornell University study found feather meal to be only 31.9% effective in preventing browsing on conifer seedlings. Go figure.

To use: Like other animal waste products that deter deer, feather meal's repellent properties are most successful in summer. Sprinkle around the garden or hang in mesh bags (the toes of nylon stockings are ideal) in branches or on posts strategically placed to protect vulnerable plants. Feather meal holds up well, and lasts 30 to 60 days before you must replace it.

Repellents

Commercial products based on egg solids include, but are most likely not limited to:

Deer Off — Faired very well in a four-year study at Rutgers University in New Jersey, repelling deer up to three times longer than 35 other substances tested. Registered by the EPA and guaranteed by the manufacturer. Apply to vegetable plants or fruit trees regularly, up to two weeks before harvest.

Deer Away — Odor/taste repellent made of 37 percent putrescent egg solids with a reported 85 to 100 percent effectiveness in field studies. Registered for use on ornamental plants, conifers, and fruit trees before they flower. Weathers well; effective for months.

Not Tonight Deer! — Uses powdered eggs and white pepper to generate a smell and taste that deer dislike. Can be used on food crops.

Pros and cons: Feather meal is a good organic fertilizer, moderately priced, and easy to handle and to use. Apply it carefully, however; an abundance of feather meal will contribute too much nitrogen to the garden, and may damage plants.

Ruffling Some Feathers

Alone, feather meal is not the food of choice for flowers or fruiting plants because nitrogen produces green growth, not flowers or fruits, but it makes an ideal dressing for lawn areas and pathways near plants that need protection from deer. Or create your own all-purpose fertilizer by mixing it with bone meal (3-15-0) and wood ash or sulfate of potash-magnesia (0-0-22). For general gardening use, mix in roughly equal parts for a 1:1:1 ratio. For vegetable gardens, use twice as much bone meal as the other two ingredients for a 1:2:1 mix.

Ammonia

Anyone who's ever cleaned a small bathroom with insufficient ventilation can attest to the olfactory and respiratory irritation ammonia can cause. The strong odor gives some people a headache. It makes a great deer repellent, however.

To use: Wearing household rubber gloves to protect your skin, soak rags in ammonia, squeeze out the excess, and tie the strips to low-hanging branches, or to stakes driven into the ground.

Pros and cons: Outdoors, the smell dissipates quickly, but take precautions when working with or handling ammonia. Do not breathe the fumes, and protect your skin from contact.

Related commercial product: ferti-lome Rabbit and Deer Repellent.

Fabric Softener Strips

One of the cheapest and easiest repellents you can try is fabric softener strips — the kind you use in your dryer. The stronger the fragrance, the better, so don't use a product that is unscented or fragrance free.

Softening Them Up

One example of how a repellent works in one setting but not another is this saga of the fabric softener.

Gardeners in New York were so encouraged by the success of hanging up strips of fabric softener to deter deer that they proudly reported their results to the horticultural newsletter *HortIdeas*. When *HortIdeas* published the results, gardeners around the country tried it.

Unfortunately, respondents in Virginia did not have the success of those in New York and wrote back to *HortIdeas* that the antistatic approach had failed them. In closing their report, they suggested that perhaps the deer in New York differed from deer in Virginia. How true. But, then, deer in the same neighborhood feed and behave differently from each other.

To use: Tie or hang fabric softener strips in or near susceptible plants. Hang them at intervals of three feet (as recommended for soap and hair).

Pros and cons: "Springtime fresh" or floral scents blend into the garden better than the smell of rotten eggs, and fabric softener strips are easy enough to install. The strips quickly become waterlogged after a rain, however, and must be replaced. Even so, they remain relatively inexpensive and are worth a try.

Processed Sewage

Here's a plus. Fertilize your garden and keep away deer in one step. An ongoing study at Cornell University (2003–2004) shows that Milorganite, a fertilizer made from processed sewage, deters deer from browsing on hostas and yews when applied around the base of these plants (rate of application: five pounds per 100 square feet).

To use: Top-dress lawns or around large plants, or work into the soil at planting time. According to the Cornell study, Milorganite should be applied once or twice a month, as well as after a snowfall. Other areas addressed by the study are how well Milorganite will perform when hung from containers in tree limbs and its effectiveness as a deterrent in winter. So far, the results look promising. Halfway through testing, deer are definitely staying away from treated areas.

Pros and cons: The good news is that using a sewage-based fertilizer reduces two chores to one. The bad news, as with many other repellents, is that it works best in areas of low to moderate deer pressure when applied at recommended rates for fertilizing. Hungry deer ignore it. More-aggressive application may make a difference in deterrent effectiveness.

Related commercial product: Milorganite.

Something Fishy

Organic gardeners have long known that if they hold their noses, they can fertilize and repel deer in one efficient step. Fish meal, fish emulsion fertilizers, and even homemade concoctions of the remnants from your last successful fishing trip make potent fertilizers that have the additional benefit of smelling revolting — to deer as well as to gardeners. The heavy, lingering scent ought to be enough to repel anything that can walk away. One homemade use of fish parts is to bury them in the soil near plants. The downside to this is that they can attract hordes of meat-eating small varmints, cats, and even larger meat eaters.

To use: Purée fish parts in a blender (diluted with two parts water for each part fish), then paint onto both sides of leaves, or strain through cheesecloth and spray on plants with a hand sprayer. (Cut the cheesecloth into strips and hang them on tree branches for added protection.) A somewhat less repulsive recipe is to mix three tablespoons of kelp and one cup of fish emulsion with three tablespoons of liquid hand soap and dilute with enough water to fill a three-gallon hand sprayer. The fish smell should keep deer too far from a plant to sample it, but if they do, the soapy taste will convince them to move on to a better-tasting yard. Alternatively, commercial products give specific directions on their use; just follow label instructions for spraying.

Pros and cons: The upside is that in areas of low to moderate deer pressure, repellents made from or that include fish by-products do keep deer away from specific plants, if not entirely out of your yard. As an extra benefit, they are great, organic fertilizers. Downsides are what to do with the blender, the smell upon application (which usually subsides below human detection levels within a few days), and possible attraction to other critters (for homemade brews or fish pieces). Adding garlic or predator scents that repel scavengers helps to overcome that side effect.

Related commercial product: DeerBusters Plant Growth Stimulant and Coast of Maine fermented salmon.

Blood Meal and Dried Blood

A "red-alert" repellent, blood meal has long been recommended as a deer deterrent, and is advocated by organic gardeners as a plant food, as well. Blood products are high in nitrogen (12 to 15 percent) and phosphorus (1.3 to 3 percent). Blood meal also contains about 1 percent potash. The message from this scent to deer is very likely one of "Others have shed their blood here — *beware.*"

To use: For shrubs or trees, make small pouches of old nylons or cloth and fill them with blood meal. The bags need not be placed nearly as close together as soap or fabric softener strips because the odor sends a more menacing message. The odor of the dried meal isn't very noticeable, but once the meal gets wet, it reeks of danger to deer. Create a barrier around your yard by hanging the pouches of meal around the perimeter, or simply place them in and around vulnerable plants. You can also top-dress beds or rows with it, which gives your plants a free lunch. Feather meal, a by-product of the poultry-processing industry, can be used in the same way.

As an alternative, reconstitute dried blood with water and spray on and around foliage.

Pros and cons: Blood meal and dried blood provide a measure of control while benefiting garden growth. The biggest shortcoming is the need for vigilance; the meal degrades and washes away, and the blood spray must be regularly reapplied. Sometimes upkeep alone makes the difference between a marginal deterrent and one that really works. Many of the commercial blood-based products, however, have incorporated binding agents to make the repellents last much longer, with excellent results. In studies, liquid applications were more effective than powders, and performed best in winter and in spring. Some commercial products are not for use on food plants.

Sometimes upkeep alone makes the difference between a marginal deterrent and one that really works.

Related commercial product: DeerBusters Deer II (for nonedible plants). Plantskydd and Repellex Deer Repellent are exempt from federal EPA pesticide registration. Plantskydd is listed by the Organic Material Review Institute (OMRI) for use in production of organic food and fiber. Advertised for use on fruit trees, vegetables, watermelon, pumpkins, cauliflower, blueberries, strawberries, grapes, soybeans, and row crops. See specific product label for use.

Be aware that blood products may attract predators, from coyotes, possums, raccoons, and rats to dogs and cats, into the garden.

Predator Urine

Among the most powerful of the "red-alert" repellents is the overwhelming odor of predator urine. The fresh markings of such natural predators as cougars, coyotes, bobcats, bears, and wolves constitute an undeniable warning that deer instinctively heed. The odors of predators that North American deer would never

Effective . . . but dangerous

You may have read about the following as viable options for repelling deer. That may be, but they also carry serious risks to human health and the environment, including groundwater contamination, the spread of disease, and the endangerment of birds, pets, and other animals.

Mothballs. Also known as naphthalene, mothballs or flakes have long been used as repellents. Gardeners report temporary success in deterring deer, squirrels, and skunks, among other creatures, with the noxious fumes. Sprinkling loose balls or flakes also increases the potential for exposure to unintended victims, such as wild birds, cats, and other pets. They are flammable, evaporate quickly (and therefore need frequent replacement), and are toxic.

Creosote. This smelly, black wood preservative, used to treat railroad ties and utility poles, has long been used as a home remedy to cause deer to turn up their noses and move on. Recently evidence that creosote contains carcinogens and contaminates groundwater supplies prompted legal action to force the EPA to address the issue. It has already been outlawed for home use in Europe.

Tankage. Many sources recommend acquiring tankage, the putrefied cast-off bits and pieces from slaughterhouses, for use in the garden. But it's a garden I'm not going to visit. While the stench of rotting flesh does keep deer away, there are serious drawbacks to its use. In rural areas the odor attracts predators, and in *any* area it draws yellow jackets and household pets. Consider also that spoiled, rotting meat and innards are loaded with bacteria, from *E. coli* to who-knows-what, none of which you need multiplying in your garden.

encounter in the wild, such as lions, tigers, hyenas, and polar bears, are just as effective. Many researchers, including those at the Connecticut State Agricultural Experiment Station in New Haven have confirmed what the deer already knew: predators scare away prey.

There's no need to stalk the woods for this natural product. Nearby zoos are the best source of predator urine — best because someone else has to collect the stuff; all the gardener has to do is bring it home and set it out. Also, you might be surprised at how many people own large cats as pets. And wolves and wolf hybrids have become an unfortunate fad.

To use: Be sure to secure a source of fresh urine. The aromatic properties break down quickly, and samples only a few days old have lost considerable potency. Store in an airtight plastic container in a cool, dark place.

Containers of urine fare better in the garden than merely a little dribble here and there. A container concentrates the odor and protects it from wind, rain, and other disturbances. Small plastic tubs that once held, for example, margarine or cottage cheese work well. Poke lots of holes along the sides to let those heady aromas waft through, and place the containers along the perimeter of the garden or near especially vulnerable plants. Do not apply predator urine directly to plants; instead, place it nearby.

Pros and cons: Predator urine is good at protecting delicate plants easily damaged by light browsing, as long as you replenish it often. For many gardeners, the most difficult aspect of using

Using urine from predators deer never naturally encounter, like lions and tigers, effectively keeps deer out of the garden.

predator urine is finding it in the first place. Also, gathering, storing, and handling urine is not for the faint of heart or nose. As a side benefit, predator scents repel a range of other creatures that feast upon your garden, from mice to moose. Predator feces work too; however, in addition to possibly contacting parasites, germs, and other deleterious things that may take more joy out of gardening than deer do, why would you want to handle the stuff?

Related commercial products: Commercial outlets for predator urine and feces are listed in the appendix.

Pleasant Aromas

Ah! Finally a breath of fresh air. Wouldn't it be lovely if something as pleasant-smelling as peppermint candy, daffodils, or an orange could deter deer from browsing? Some enterprising entrepreneurs thought so too, and came up with concoctions that reduce deer damage while delighting the (human) sense of smell. Think there has to be a catch? Well, sort of. Like other scent repellents, products using these scents are aimed at offending deer, but unlike the others, they are also meant to be pleasant to humans. The catch is, the pleasant aromas are not necessarily what deters the deer. While peppermint (and other strongly scented mints or aromatic plants) are known to repel deer simply because of the masking ability of the intense scent, the product manufacturer that takes advantage of this (see below) also reinforced the scent with white pepper and garlic oil. Likewise the orange-scented soaps. Though the orange fragrance may be strong enough to mask odors, the ingredients in the soap base are good insurance that deer will find the overall product unappealing. However, the daffodil scent stands on its own. Since deer don't like daffodils, a garden that smells as though it is exclusively planted in those unappetizing flowers, is one deer ignore.

To use: Hang up or spray on according to manufacturer's instructions for application.

Pros and cons: The products listed below are easy and pleasant to apply, have provided good success for residential users in premarketing trials, and smell nice! All of the products listed here are approved for use on food crops.

Related commercial products: Deer Out (based on peppermint oil, white pepper, and garlic oil), Deer Solution (cloves and cinnamon), Deer Chaser and Deer No-No (citrus-scented soaps that come as solid blocks in mesh bags, ready for hanging).

Bad Tastes

Taste repellents work differently from odor repellents. Rather than forming an odor barrier to an area, taste repellents, also called contact repellents, protect the exact plant or leaf to which you spray, brush, or otherwise apply them; deer must taste the repellent before it can take effect. This is both their greatest success and their biggest drawback. If any other food is available, deer can't stomach leaves or stems coated with such nasty stuff, but first they must learn. Unfortunately, learning means tasting and tasting means at least a little bit of damage. Most deer will try several bites before they realize that the entire plant (or border or garden) tastes awful.

Plants you have treated to taste bad to deer won't taste good to you, either.

For gardeners who enjoy deer in or near their yards, contact repellents may be the perfect answer. Deer do not withdraw from the entire area; they may still meander through now and then, even once they decide you have nothing in your yard worth eating. But unless starvation threatens, your bitter bushes and sour grapes will not tempt them to taste again.

Just as area repellents have cons as well as pros, contact repellents have a few shortcomings of their own. First, most taste repellents are meant exclusively for nonfood plants: Plants you

Related commercial products: Miller's Hot Sauce Animal Repellent is labeled for use on shrubs, Christmas trees, fruit trees, and vegetables. You must apply it before the fruit set or vegetables develop. It's expensive, but so concentrated that eight ounces will cover an acre. Green Screen (meat meal mixed with capsicum), Hot Pepper Wax Animal Repellent, and N.I.M.B.Y. (hot pepper mixed with castor oil to combine a nasty taste with a sick tummy) are other examples.

Green Eggs and . . .

That's it. Putrid eggs. The same stuff that smells terrible actually tastes worse. So while you're concocting Eau de Rotten Eggs (see page 111) as an area repellent, make enough to cover thoroughly any particularly vulnerable plants.

To use: Spray, drizzle, brush, or splatter over plants to be protected. If the smell doesn't keep deer away, the taste will.

Pros and cons: A double-duty deterrent, rotten-egg mix very effectively dissuades deer. This has been proved over and again by gardeners and in independent, scientifically controlled testing. A Colorado State University study found a ratio of one part water to four parts egg (liquefied) to be highly effective in deterring deer and moderately so in discouraging elk. The smell becomes unnoticeable (to people) once the solution dries on foliage, but the taste remains. Rotten eggs should not be used on any plant meant for human consumption.

Related commercial products: Deer Away and Deer Off.

Soap Spray

Soapy water, especially when laced with human scent, serves as another double deterrent to deer. It not only smells bad, but also tastes terrible to deer, according to a Wisconsin gardener who suggested this solution to the gray-water-disposal problem.

To use: Save soapy bathwater or whip up a fresh batch of soapy water from bar soap. Spray on the foliage of vulnerable plants. Because some plants are sensitive to soap, be sure to test-spray a patch before treating fully. Detergents are usually synthetic versions of soap, lacking the organic components (animal-based fatty acids) that repel deer, so don't expect detergents to affect deer the same way.

Pros and cons: Though his approach has not been thoroughly tested, the reporting Wisconsin gardener claims great success using his sudsy solution. Gray water offers the additional advantage of being free, the basis is reasonable (deer are repelled by soaps, and if you've ever had your mouth washed out with it, you know it tastes bad). One drawback is the fact that you have to spray again after a good rain or two. Also, soap leaves an unattractive whitish film on foliage and can attract rodents. If mice or similar varmints are a problem in your garden, stir in a tablespoon of hot pepper sauce.

Related commercial product: Hinder.

Thiram

Originally developed as a seed protectant, the fungicide thiram has proved quite distasteful to deer. It irritates mucous membranes in and around the mouth and nostrils. Be sure to check the label and follow the instructions regarding which plants this product can be used on. (Never apply any commercial products to food crops unless they are specifically labeled as approved for those crops.) Thiram is most often used on dormant trees and shrubs. Although it does not weather well, adhesives such as Vapor Gard can be added to increase its resistance to weathering. Thiram-based repellents also protect trees against rabbit and mouse damage.

To use: Mix one part thiram with one to two parts water, and spray on susceptible plants. Be sure to spray up to a height of at least six feet.

Pros and cons: Although apparently very effective in reducing deer damage, thiram has several drawbacks. First, you should use it only on plants for which it is registered. Second, it can be applied only to dormant plants, and only when temperatures are above freezing. It also tends to be expensive.

Thiram is considered a harsh chemical. Avoid all contact. It can cause confusion, cough, dizziness, and eye and skin irritation, and is suspected by the Centers for Disease Control and Prevention to cause birth defects; pregnant women should avoid using it. Thiram is not approved for organic gardening.

Related commercial products: You can purchase thiram in straight form as a fungicide and in animal repellent concentrations, as Nott's Chew-Not, Chaperone Rabbit and Deer Repellent, Deer Ban Deer Repellent, and Gustafson 42-S.

Bitter Chemicals

Suffice it to say that if you saturate plant parts with mouth-biting, stomach-sickening, bitter-tasting chemicals, deer will avoid them if at all possible. Home remedies using chemicals are absolutely discouraged, and commercial applications are not for use on food plants. Only one commercial product, Ro-Pel, has been recommended as a topical (spray on) application against deer, though others may follow. Ro-Pel is for use on everything from ornamental plants to wicker furniture, to deter all sorts of gnawing creatures, including deer. Use full strength to prevent chewing by deer. A quart covers from 1,000 to 4,000 square feet.

To use: Carefully follow manufacturers' instructions. Avoid breathing the spray. It can be used to protect nursery stock, annuals, perennials, and shrubs.

Pros and cons: Though effective in many uses against chewing animals, this treatment hasn't always fared well in deer trials, though independent reports vary. There is no odor, but the taste is reputedly awful. It is reported to work well on flowers or plants after animals have begun to chew them. They have to taste the product to learn to avoid it. You may need to reapply this treatment following rains or watering, which could account for the varying success rates. Once animals learn that the taste is bad, even a trace of the chemical is enough to remind them. May be poisonous to some plants.

Related commercial products: Ro-Pel spray. (Benzyldiethyl [(2,6 xylylcarbamoyl) methyl] ammonium saccharide)

Systemic Aversives

Systemic aversives, for lack of a simpler term, are substances absorbed through the roots of plants to make them taste bad through and through. The best example, denatonium benzoate — known as Bitrex — is widely regarded as the most bitter substance around. As little as 30 parts per million is too bitter for humans to tolerate. Once absorbed through the roots, the entire plant becomes intolerably bitter.

To use: Follow manufacturer's instructions. Bury one denatonium benzoate tablet so that it contacts the roots of each plant that needs protection, and the plant's "circulation" system will do the rest. The plant's roots absorb the chemical and distribute it throughout the plant, making all parts bitter tasting. These products take from 30 to 45 days to take full effect and work best during periods of active plant growth. Though they will protect dormant plants the winter after they have been absorbed into the plant's system, these tablets cannot be absorbed when the plant is dormant. A single tablet should last at least a year. Liquid products are absorbed through the leaves or roots.

Pros and cons: This product is simple to use, safe, and nontoxic; however, denatonium benzoate tablets are not inexpensive compared to other repellents. As with other taste repellents, the deer must sample the forbidden fruit before they learn the consequences. Though wonderful for ornamentals, this is obviously not an option for food crops. What tastes bad to deer will taste bad to you or other animals. The time delay between treatment and effectiveness means you had better have a contingency plan in the meantime. These products are reported effective against all chewing animals, woodpeckers, and certain insects, including beetles.

Related commercial products: Products based on Bitrex include DeerBusters Deer Repellent Systemic Tablets, Durapel, Repellex, Repellex Systemic ML2, Tree Guard, and Ro-Pel. Other systemics are Mirrepel's Gold'n Gro Guardian, which is applied as a foliar feeding spray or root drench.

One-Two Combos and Triple Threats

Knowing that some things deter some deer in some places, some of the time, gardeners, researchers, and commercial entities use that knowledge to create new deterrents. By combining repellents, they can create a product that is superior to any of its individual ingredients. Although it is still more effective to rotate and change various repellents from time to time, the array of possible combinations makes keeping one step ahead of browsing deer that much more plausible.

Think of combining ingredients as launching a back-up plan right along with your main mode of defense. By adding taste (contact repellents) to scent (area repellents), you cover your bases two ways. If a deer ignores or adapts to the smell, the bad taste should send him on his way. Employ separate deterrents individually (for instance, hang bars of soap, spray with egg mixture, and work blood meal into the soil) or combine them into one "end product."

This is where your imagination is truly the only limit. Consider combining hair and blood meal or feather meal and hanging it in pouches, adding cayenne to your stinky egg spray or fish emulsion fertilizer, lacing soap spray with peppermint oil, or any other combination of the individual raw ingredients listed above. Commercial products that use this strategy are Deer Blocker, Master Gardening Deer Repellent, Bonide Shot Gun Deer Repellent, and DeerBusters Deer I Repellent (which combines eggs and garlic with capsaicin), Not Tonight Deer! and Deer Out (eggs and white pepper), Deer Off and Havahart Bulb Guard (egg and capsaicin), N.I.M.B.Y. (which combines the sickening agent castor oil with the painful bite of capsaicin) and Bobbex Deer Repellent (Garlic oil, cloves, fish meal, fish oil, onions, eggs, vanillin, and wintergreen oil). See product appendix for specifics.

Keep 'Em Guessing

Deer are creatures of habit, and one way to keep them out of the garden is to continually disrupt their routines. One way to do this is to "mix up and freshen up" your repellents. For example, both hair and blood products work well; they work even better when mixed together. Keep the scent stronger by freshening the mix every couple of weeks.

Startling Sights

No one has reported success keeping deer away for more than a few days using only visual deterrents, but in combination with other deterrents they can make an important contribution. In general, two or more senses always confirm a threat much more convincingly than can a mere single-sense suspicion.

Anything visually new or strange will make deer instantly suspicious, but they get over it. Deer have what is known as a "flight distance." If something comes too close, they take flight. Oddly enough, however, if they don't notice something until it is already within their flight distance, they may ignore it. It's as if they think, "Well, if you've gotten this close you must be harmless — otherwise, I'd be gone by now."

This is why an old-fashioned scarecrow doesn't work very well on deer. They can't recognize a scarecrow as a human form much beyond a hundred paces, and once they get that close, they may decide that it poses no threat. After the deer have seen the scarecrow a few times, it becomes part of the scenery. You can try moving a scarecrow every few days and dressing it in floppy clothes that wave in the breeze, but it still won't be terribly effective. But a visual deterrent that actively moves is another story. If your scarecrow suddenly jumps or waves its arms, the deer depart posthaste. Menacing movements make the difference.

Mechanical Gizmos

Whether you wind them up, pop in batteries, or plug them in, there are so many moving mechanical gizmos out there, from toy robots to overhead fans, that the only limits are what you have available (or how much you're willing to spend) and your imagination. Anything that suddenly moves will startle deer, especially if it was sitting there minding its own business before it jumped. If it advances in their direction, so much the better. When a predator scent or strange noise accompanies the beastie, the result is an effective scare.

To use: Be ingenious. With a little imagination, any gardener can create, for example, his or her own version of Frankenstein's monster to stand sentry. Give it some erratically moving parts and watch the deer stand back. One gardener attached "arms" to a post with hinges and fed heavy-duty fishing lines from the ends of the "arms" up through an eyebolt in the head. He dressed the scarecrow in his own dirty, smelly clothes. The fishing line went along the garden fence, fed through a series of two or three more eyebolts, and ended up on the gardener's porch some 80 feet away. When the gardener spotted birds, critters, or deer in the garden,

Deer have what is known as a "flight distance." If something comes too close, they take flight. But if something is already within their flight distance when they detect it, they may ignore it he yanked on the fishing line and the scarecrow started waving.

That may just be the tip of the iceberg of creative movable-scarecrow ideas. Try breeze-waving reflective bird deterrent tape or surveyor's tape and a string of rattling tin or aluminum cans and dangle them from his outstretched arms. Rig his arms or legs to move in the breeze or by such low-tech means as pulling an attached string. One innovative gardener in Texas reports a scarecrow that has been successfully standing guard duty for seven months. Her secret? She scented the scarecrows' clothes with deodorant soap and attached movable eyes.

A scarecrow with movable parts can be an effective deterrent.

Pros and cons: Any moving mechanical gizmo will probably frighten deer from your yard, at least initially. But if the deer learn that the thing can't catch them, they won't run away from it. Your best hope may be that they won't return after the first encounter. Deer's tendency to disregard what doesn't chase them is a prime reason why rotating, moving, and otherwise changing deterrents is so critical. A string of tin pans glinting and rattling in the breeze "over there" is one thing, but to the deer, it's a whole other threat when it (apparently on its own) moves to another spot in the yard.

Related commercial products: Mechanical scarecrows sold for orchard protection are available in several models. One of the best is the Scarey Man scarecrow, which employs a battery-powered light, siren, and inflatable body to freak out deer that come near. Different models can be programmed to be active either day or night (or both). The balloonlike body suddenly inflates and the siren screams at set intervals.

White Flags

Grazing peacefully at meadow's edge, the group of deer suddenly tenses. The atmosphere changes in an instant from one of pastoral peace to adrenaline-charged anxiety. One doe stamps a foot, another turns toward the woods. When a third raises aloft that white flag, all bound for cover. What was it? Who knows? But whatever the source of the scare, the message was in that white flag.

To use: Fashion flags from white cloth or rags, or recycle those flimsy white plastic grocery bags into a flag that will fill with air (like a windsock) and wave all on its own in even the slightest breeze. Flags about ten inches long and five or six inches wide mimic those of the real McCoy. Position the phony tails at strategic locations around the garden. Tie them to posts, branches, or fencing so that they will wave in the breeze.

Pros and cons: White flags are meaningless to blacktails and mule deer, and the best that you can hope for is that they will make visiting whitetails nervous. Used with additional deterrents, white flags may confirm a nervous deer's suspicions that danger lurks nearby. If you want to try the fishing-line technique, consider purchasing a monitoring device that will alert you when something has entered your yard.

Floodlights

Neither gardeners nor wildlife biologists seem to be able to agree on whether lighting makes a difference to dining deer. After all, it's not ambience they seek.

You can time floodlights to go on at irregular intervals or connect them to a motion detector.

To use: Install floodlights so that the beam shines into the yard. Plug them in and see what happens.

Pros and cons: Motion-sensor lights rank high on the list of things insurance companies like to see in your yard anyway, so for the sake of home security alone they are a good investment. Although relatively inexpensive and easy to install and operate, they lose their effectiveness as a deterrent as soon as the deer figure out that nothing changes except the lighting.

Related commercial product: Blast Away Deer solar floodlight.

Offensive Sounds

Imagine being deep in the forest, the ground soft and scented of humus. Raindrops fall, leaves rustle in the breeze. Birds, mice, and squirrels call and scamper. You are enveloped in small, muted sounds and thick, familiar scents. Between heartbeats, something changes. A tick of a watch, the crunch of brittle leaves a hundred yards away, and you're gone. Classic behavior. If you're a deer.

A deer's sense of hearing is vital to its every action. Watch those radarlike ears swivel and then suddenly focus, and it's obvious that they're listening to everything around them. Having learned which sounds are "normal" in their world, deer filter them out, but they instantly detect and quickly evaluate the smallest "wrong" sound. They may stare in the direction of the noise, ears cocked full forward. They may stamp a forefoot or advance toward the sound in an effort to force any hiding predator to betray its whereabouts. Or they may immediately flee for safety.

Having learned which sounds are "normal" in their world, deer filter them out, but they instantly detect and quickly evaluate the smallest "wrong" sound.

In urban and suburban areas, deer have long realized that we humans are a noisy lot, and frightening them away with sound may not be as easy as in quiet country settings. The deer must perceive the sound as a genuine threat.

Shake, Rattle, and Roll

Like most of the best deterrents, those that shake, rattle, and roll actually assault more than one sense — in this case sight and hearing. Pie pans, planks, and strings of tin cans have been a staple in the gardener's arsenal for generations.

To use: There are limitless possibilities here. The idea is to hang, by twine or invisible monofilament, something that moves and rattles in the breeze. Examples include:

- *Aluminum pie pans:* String them over susceptible plants in pairs. As the pie pans move with the wind, not only do they clank together, but the motion and sporadic glare from the shiny surfaces also create a visual deterrent.
- *Tin cans on a string:* Tie tin cans in bunches and hang these clattering wind chimes about the garden.

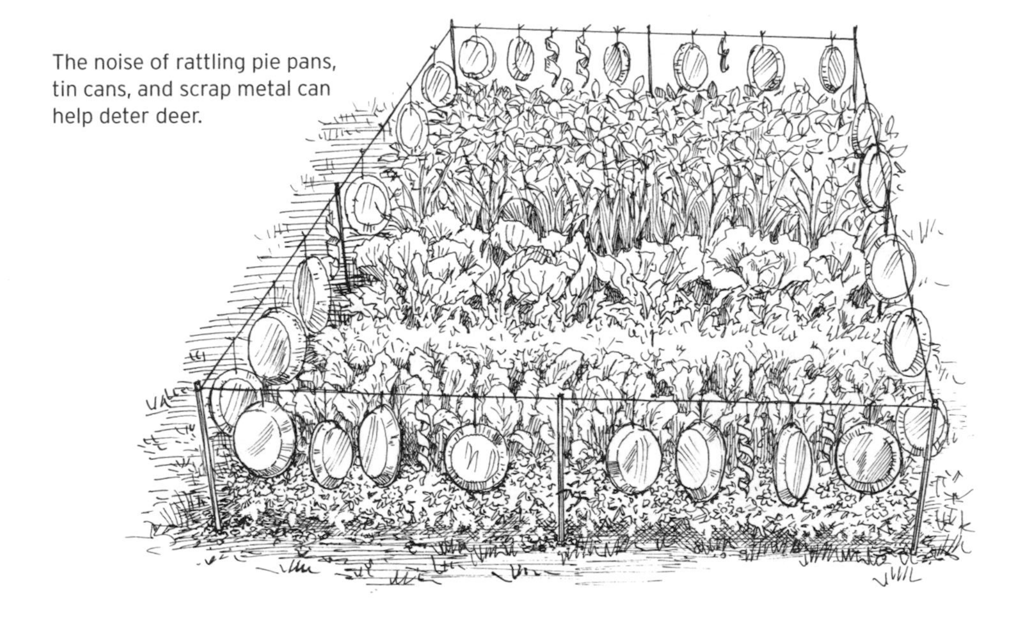

The noise of rattling pie pans, tin cans, and scrap metal can help deter deer.

- *Scraps of metal roofing, fiberglass siding, or old planks:* Drill a hole at one end of a scrap about six inches wide and three feet long. Thread heavy twine or rope through and hang in a tree or other spot where the scrap will bump and rattle with the wind. Every few days, move the deterrent to a fresh location.

Pros and cons: These are time-honored, inexpensive deer deterrents that make good use of materials on hand. Such devices do fend off deer, for a while, but are somewhat labor intensive to set up and move around. As the season wears on, the deer will get used to them, so take them down entirely and employ a different set of deterrents, then reinstall them.

Radios

Whether it's talk radio or hard rock, the sounds carried by radio waves are not soothing to deer. Deer rely heavily on those large, swiveling ears to keep them filled in on the goings on around them, and the noise of a radio drowns out their first line of natural defense. Thus, until they get used to them, deer find those sounds very unnerving. But therein lies the rub, for get used to them they will, as evidenced by the fact that deer can be seen congregating next to loud, busy roadways to feed.

To use: Some people recommend tuning the radio between stations to make the sounds as erratic, high-pitched, and annoying as possible. Deer are sure to notice the strange sounds, but that does not necessarily mean they will interpret the noise as dangerous. Other folks swear that talk radio, especially a heated debate, with rising and falling emotional human voices, offers the best deterrent. Still others claim that deer find more frightening the crashing crescendos of classical music or the violent beats of hard rock.

Tune the radio as you see fit, turn up the volume, and place it in the garden before dusk. Let it play all night or, if it has an alarm, set it to begin playing just before peak deer activity. Protect the radio from the weather by placing it under an overhang or inside a waterproof container with at least one open side — a five-gallon plastic bucket turned on its side will suffice. Move the radio around from night to night to prolong the period of effectiveness. Changing stations or frequencies periodically may help.

Pros and cons: Strange sounds and/or human voices emanating from the greenery are sure to disturb deer, but probably not for long. And who else will the radio disturb? Neither you nor your neighbors should have to lose sleep over deer in your yard. If, however, you can play the radio loud in good conscience, by all means add it to your repertoire. It's clean, it doesn't smell bad, and it just might keep the deer away from the garden for a few more days.

Things That Go Boom

Nothing beats a sudden loud blast for sending deer scattering from the scene in a hurry. Propane or gas cannons and sonic bombs work well. They generally protect an area of from five to ten acres, making them most useful for commercial sites, such as nursery tree plantings, orchards, and large crops.

To use: Be sure to follow the manufacturer's guidelines regarding fuel, setup, and maintenance. Position the noisemaker near the center of the area that needs protection and set it to go off at desired intervals. Regularly alter the timing between blasts to prolong the effectiveness. Most commonly used to protect orchards, row crops, and truck crops, exploders work best when they go off at random intervals or rotate the direction of their blast (called a Double John). Changing the location of the cannon and varying the rate of fire will extend the usefulness of cannons as deer deterrents.

Pros and cons: Your neighbors may never speak to you again, unless they too suffer from unbearable deer damage. Before resorting to such a measure, meet with any neighbors within cannon shot and discuss your concerns. If they share your deer problem, they may even offer to foot some of the bill, as the exploders can run into the hundreds of dollars. (Wildlife agencies sometimes have them for rent or loan to private individuals.) The booms will scare away birds as well as deer, and probably other wildlife, but deer become accustomed to the noise in about a month.

Related commercial products: High-pitched sonic warning devices can also be purchased from various outlets.

Ultrasound

High-tech comes to gardening. A variety of ultrasonic devices are available. According to the manufacturers, deer (and other animals) find high-pitched frequencies intolerable. Because, by definition, humans can't hear ultrasound, it offers the benefit of annoying only the animals. As delightful as this premise might seem, there's just one problem. According to the few valid studies that have been done on deer hearing (or specifically on the effects on deer of ultrasonic noise) these devices don't work — at least not on deer.

You may recall that deer have been tested to hear best in the two to six kilohertz (kHz) range. Ultrasound includes sound waves higher than 20 kHz, well beyond that of a deer's range of hearing. A controlled study of confined white-tailed deer failed to elicit any noticeable response to the sound in this range. Researchers documenting every flick of the tail, twitch of the nose, and flicker of the ear could find no discernible response to ultrasonic sounds.

One manufacturer, asked to provide test results that might at least imply that these devices had any effect on deer, suggested that consumers buy one of each type and decide for themselves. Bottom line: many other types of deterrents have much better proof of effectiveness.

gets watered. Most reports on effectiveness, so far, are anecdotal. Many gardeners report delight, if not downright glee, at hearing the *psssst* blast in the wee morning hours, followed by a quick retreat of hooves. Manufacturers claim that the full-hose blast of the Scarecrow model deters not only deer, but also dogs and other animals from protected areas. They add the comforting thought

The No-Fence Dog Fence

Hate the chain, but not crazy about fencing in your yard? A great idea for gardeners is the no-fence dog fence, sold under such brand names as Invisible Fence and DogWatch. These are electronic containment systems that train your dog not to cross an "invisible line." A thin cable is buried around the perimeter of the area to which your dog will have access. The dog wears a radio collar that transmits a signal to warn him whenever he ventures near the line. If he gets too close, *zzzaapp!* Wouldn't it be nice if we could just get these on the deer?

After a brief training period, most dogs learn the limits and stay within their "territory," although some will ignore the shocks if the incentive – giving chase to a rabbit or, unfortunately, a deer – is strong enough. The breeding, training, and individual personality of your dog are the deciding factors. If an electronic fence succeeds in restraining your dog, you can be sure that deer (and other sneak thieves) will give the loose canine, and your garden, a wide berth.

One study published in the *Wildlife Society Bulletin* compared dogs to chemical repellents for effectiveness in keeping deer from browsing in a forest restoration area. In the study, dogs (of undisclosed pedigree) were kept in the area with a wireless fence system for three years. Browse rates were 13 percent in the dog-protected areas, 37 percent in the areas protected with chemical repellents, and 56 percent in the control area, where no repellent was used.

Another study at Cornell University tested the Off-Limits system, which also pitted electronically fenced dogs against browsing deer, this time in a two-year study in commercial orchards. In the second year of the study, bud loss was 25.2 percent in control areas, with nearly three times as many flower clusters produced in the dog-protected areas than in the control areas, and twice the ultimate yield. Because they were protected from browse damage, protected trees were also 61 percent taller than the control trees.

that the water blaster can't tell a deer from a burglar, but will startle and send any and all unwanted visitors away from the area they protect — soggier and wiser.

Man(kind)'s Best Friend

Unsure of just how much rotten egg or coyote urine you want in your yard? Get a dog, instead. Deer avoid dogs, even the little yippy ones. And for good reason. Dogs have replaced wolves, cougars, and bears as the number one nonhuman predator of deer. Especially in winter when deep or crusted snow makes it difficult for deer to outrun them, dogs exact a serious toll. They still have that canine predator instinct, and the deer know it.

Not that having a guard dog for your garden doesn't carry its own costs and responsibilities. To begin with, very few areas allow free-roaming dogs anymore. Loose dogs often band together, or pack. When they do, it's as if ancient memories of their wolf past suddenly revive, and family pets revert to wild-eyed, dangerous beasts. Dogs in packs attack livestock, other pets, wildlife, and people. Things your gentle pet would never do its alter-in-pack-ego does without a second thought. Many areas allow people to shoot on sight dogs that are menacing or chasing livestock or deer. Therefore, for the animal's safety, that of the neighborhood, and your own peace of mind, the first consideration of dog ownership is keeping the dog under control. For that, you have three basic options: chaining, fencing, and electronic containment systems.

Chaining vs. Fencing

Life at the end of a chain is a miserable existence. And as far as protecting the garden goes, deer will quickly realize that the dog can only go so far and will then ignore him. After a while, the dog will realize that the deer are ignoring him and thus no longer fun to bark at. Stories abound of dogs sleeping while deer browse nearby. One solution is to move the chain from place to place every few days, which will help to keep the deer confused.

Before deciding to chain your dog, also consider this: If deer have found their way to your yard, what else has? Other wildlife, such as coyotes, have become just as adapted to the presence of humans, and a dog at the end of a chain is at their mercy. In some areas larger predators, such as cougars and bears, have followed their prey, the deer, and a chained dog has no hope against such an adversary. Other dogs allowed to run loose have ganged up against their chained counterparts, with sickening results. *So don't chain your dog.*

Instead, fence. The wonderful thing about fencing a garden or yard with a dog inside is that you needn't worry so much about fencing the *deer out* as fencing the *dog in*. Which is a darned sight easier to do.

The height and sturdiness of a dog-proof fence depends on the height and sturdiness of the dog. Medium-sized dogs can jump over a three-foot-tall fence, but generally won't attempt it if they don't have to. A four-foot fence of wire mesh, plastic mesh, or wood, supplemented with a strand or two of electrified wire, will keep in almost any dog. Choose a standard chain-link fence, a charming white-painted picket fence, or any other style that suits your fancy. A dog fence is ideal for many people because it allows the dog free run of the yard and garden, which means he gets lots of attention from his people and can serve as an alarm system for everything from deer to burglars. As part of a security system for your home, the fence may even add value to your property.

A double-layer fence (in which one fence is constructed three to four feet outside of another fence to discourage deer from trying to jump) is a great setup for a guard dog. Simply use the area between the two fence lines as a dog run and you have added the threat of predation to the intimidation of a double fence. This kind of fencing also keeps the dog from trampling your turnips.

Choosing a Dog

Do you need to invest in a purebred, guard-trained Doberman? Of course not. Any dog that looks like a dog, smells like a dog, and barks and runs around like a dog is dog enough to make deer nervous. Again, those ancient wolf-pack memories may be to blame. Perhaps the deer see, smell, and hear *dog,* but register *wolf.*

In general, dogs bred for guarding, hunting, or herding make the best "deer dogs." Such guardians as Dobermans (my favorite), Great Danes, German shepherds, Komondors, and Great Pyrenees top the list. Hunting breeds, especially hounds, will sound the alarm when deer come around, but trained hunting dogs are taught to ignore deer scent as "trash" because it distracts them from the hunter's preferred prey. Herding dogs, such as Border collies, collies, and the many types of shepherds, have the instinct to chase and may or may not be inclined to bark at intruders.

Before investing in a purebred, you may want to take a trip to the pound. First and foremost, look for the same things you would look for in any pet dog. A dog that seems happy to see you is a good start. But if he's bounding around uncontrollably in the kennel, he's apt to behave the same way in your garden. Pick a dog you want to live with. Second, look for a dog that exhibits some of the characteristics of the breeds mentioned above. Most animal shelters will try to venture a guess at the breed (or breeds) a prospect is made of. Third, look for a bright-eyed, alert dog of medium to large size that will look you in the eye and respond to changing tones of voice.

Despite the old shepherd's saying that if you have one dog, you'll get the work of one dog, but if you have two, you'll get the work of half a dog, getting two or more dogs to work as deer guards is a different story. Remember the old "pack instincts." Dogs tend to home in on "prey" harking back to their wolf ancestors' hunting days, much better when there are two or more of them. This is one instance when the pack mentality works in your favor. Dogs tend to be more alert, more aggressive, and noisier when working together than when working solo. It's almost as if they're trying to impress one another. And all of the studies known to date have employed more than one dog.

Tactics NOT to Try

Over the years, a lot of desperate gardeners have tried a lot of desperate means to keep deer away. The worse the situation, the more drastic the tactics employed. But there are limits to what we could and should try to alleviate a landscape or gardening problem.

In some cases, the limits have been defined by law. In others, the experiences of those who have gone before us, or plain old common sense, should tell us we have crossed the line. If your deer problem is so intense that you would seriously consider such an extreme tactic as poisoning or shooting, then the situation has mushroomed beyond a gardening concern and is an issue of wildlife management and public policy (see chapter 7, "Community Efforts"). Consider the following discussion of what are considered "bad ideas" and why they are not recommended.

Stinker Repellents

One so-called clever idea is that using human urine or feces will keep away deer. It could be true, but if you try it, please don't invite me over. Although urine is sterile and loaded with nitrogen, the smell is drawback enough for most people. Human waste left out in the garden violates public health laws and is incredibly unsanitary. Again, germs abound and you don't want to give them free access to your garden.

Poisoning

Stated simply, don't. *No* poisons are registered in the United States for deer control. Poisoning deer with any substance for any reason is illegal. No matter how big a pest a deer has become in your yard, death by poisoning is a hideous, undeserved end.

Trapping

Wildlife officials sometimes trap and remove nuisance deer, usually to relocate them. Such devices as rocket nets, drop-door box traps, and tranquilizer guns are used to catch deer with as little trauma as possible. Officials then transport the captured animals to an area of lower deer population.

Unfortunately, research reveals that relocated deer do not fare well. In one study, only two of 15 tagged and relocated deer survived a full year. Even in a recent attempt to reestablish a population of the rare Columbian whitetail, in which extreme care was taken due to their endangered status and veterinarians were on hand, five out of 24 deer rounded up died. Add to that the exorbitant cost of capturing and moving deer, and the consensus is that the result seldom justifies live removal.

Traps set up to kill deer are illegal.

Unpermitted Shooting

All states protect deer under the law as a game species, and will prosecute anyone shooting deer illegally. However, most states allow farmers and landowners to protect their crops or property from wildlife damage. Wildlife managers concede that the most effective means to stop deer that are in the habit of raiding a garden, field, or orchard is to remove the offending deer. Shooting, or in some cases bow hunting, is the most efficient means. Permits may be issued that allow a farmer or landowner (or in some cases his appointed agent or other hunter) to kill deer that are causing damage. Requirements differ from state to state, with mixed results. Stories of farmers herding deer into traps and shooting them by the dozens show how drastic the problem of, and the solution to, deer depredation can be.

In order to obtain a crop damage permit, out-of-season kill permit, or deer depredation permit (names vary by region), the first step is to call your local department of wildlife or natural resources. It may or may not be the regulating agency in your state, but people there can certainly route you to the proper authorities. In Rhode Island, for instance, the Division of Agriculture controls deer damage permits. Under the law in Massachusetts (Massachusetts General Laws chapter 131, section 37), no permit is required to destroy deer causing crop or property damage, though there are limitations as to how the landowner can kill deer, and the kills must be surrendered to the state environmental police. In many states, including Wisconsin, Indiana, Ohio, Maryland, New York, and Virginia, you will need to show proof of deer damage, and in some states, such as Wisconsin and Indiana, the damage must be in excess of $1,000 per year in order to qualify for the permit. In Virginia, you need only the game warden's written permission to kill deer where damage occurs. So call before you clean up proof of deer damage!

How you implement a permit also varies, but most states allow shooting only antlerless deer (which also varies somewhat by definition) and regulate what type of weapons can be used. Generally, nothing under .22 caliber is allowed, nor any semiautomatic weapons. Some states limit the number of deer that can be killed; others, such as New Jersey, do not. And not all depredation permits are for rural settings. Indiana, for example, has an urban deer policy that allows bow hunters to remove antlerless deer in predetermined urban zones. Of course, these laws are ever-changing in response to public outcry and commercial farming pressure, so some of them could be different before the ink dries on this page. The bottom line is that if you have significant enough deer damage on your property to consider shooting the animals, call your local authorities to find out what your rights are.

Comparing Deterrents

Deterrent	Longevity*	Relative Cost	Comments
REPELLENTS			
Soap	Several weeks	Low per individual tree, moderate per square foot	Very easy to use
Hair	Few weeks	Free	Very easy to use
Repellent plants	Indefinite	Low to moderate	Avoid toxic plants
Garlic, rotten eggs	Few weeks	Free to low	Smells bad during preparation; application easy to messy
Fabric softener	Until hard rain	Very easy to use	Very easy to use
Processed sewage	Several weeks	Moderate	Very easy to use; commercial products may contain heavy metals
Blood meal	Days to few weeks	Moderate	May attract predators
Predator urine	Days to weeks	Free to moderate	Challenging to acquire
Hot pepper spray	Days to weeks	Free to low	Don't apply to food plants within one week of harvest
Egg spray	Days to weeks	Free to low	Requires preparation
Soap spray	Days to few weeks	Free to low	Easy to use
Thiram	Weeks	Moderate	Follow product label
Systemic aversives	Indefinitely	Moderate to expensive	Don't use on food crops
OTHER DETERRENTS			
White flags	Weeks	Low	Work only for whitetails
Mechanical gizmos	Weeks	Free to moderate	Easy to use; may require assembly
Floodlights	Weeks	Moderate to expensive	Require installation
Homemade noisemakers	Weeks	Free to moderate	Require assembly
Radios	Weeks	Moderate to expensive	Very easy to use
Propane cannons, sonic bombs, sonic warning devices	Weeks	Expensive	Require installation and tolerant neighbors
Ultrasound	Indefinite	Moderate to expensive	Questionable effectiveness
Monofilament	Days to weeks	Low	Watch where you walk
Timed sprinklers	Indefinite	Moderate to expensive	Remember when they start!

* Longevity is relative to many factors — especially, for repellents, weather conditions. In general, repellents with a longevity of "several weeks" will last somewhat longer than those with longevities of "few weeks," conditions being equal. For other deterrents, longevity is largely a function of how quickly the deer adapt to the particular technique.

SIX

The Deerproof Garden

CONVENTIONAL WISDOM HAS LONG HELD that the *only* way to keep out deer is with fencing. And not just any fencing, mind you, but the type that costs hundreds of dollars and looks like a prison barricade. My garden in Elk, Washington, had such a fence — 8 to 10 feet high in places, constructed of pine poles and woven sheep wire, and topped with two strands of barbed wire. At least one deer still got in. It probably crawled under a gap in the wire.

Although such fences usually work to keep deer on one side and your yard on the other, these days you are not nearly so limited to such conspicuous materials. The options available today make it much easier for homeowners to build a barrier that is both deerproof and attractive. Fortunately, deer fencing comes in so many types, materials, and varieties that at least some of these options will be just right for your situation.

Fencing Lessons

For areas of high deer pressure, wildlife experts across the board still recommend fencing as the most reliable method to ensure against damage in home gardens. And because many gardeners in many situations and with many different means continually come up with ways to keep deer out, you will find that there is an ingenious array of designs and fencing materials that do just that. Some fence designs rely on sheer height, others combine elements of height and width, and a few even add the zing of electricity. One type of fence lures deer to the edge of forbidden territory only to deliver a shock that teaches them to avoid the area. But by far the most attractive options in deer fencing for the home garden are invisible. Deer netting, strong enough to hold back a running deer, yet designed to melt visually into the background, has become extremely popular over the past several years as a variety of products have become available.

Fence First, Plant Later

Negotiating fences is a learned behavior for deer. They won't bother trying to get through a fence if it is put up before there is anything planted inside of it. Erecting the fence first and planting later is one more way to teach deer to look elsewhere for food.

What Makes Deer Nearly Unstoppable

To understand what kind of fencing it takes to really keep deer out, their physical strengths, which are many, and their limitations, which are few, must be considered. Any healthy deer, regardless of species, can clear a five-foot fence without even trying. Most can vault six or seven feet, and some can breeze over an eight-footer. When panicked or under enough stress, whitetails have been observed jumping in excess of nine feet high. Smaller than most species of whitetail, the black-tailed deer, common to the West Coast, are not as accomplished at jumping, and most cannot clear much over six feet. But blacktails are even sneakier at finding gaps or small spaces to squiggle through than most of their white-tailed cousins. One study reported that a nine-inch-square gap is all it takes for a blacktail, or a small whitetail, to get through a fence. Amazingly agile for their size, elk have been known to jump more than eight feet high. And moose can step over a four-foot fence without missing a beat. Generally, however, they don't bother trying to get over what they can just go through. Deer can also cover a wide expanse in each stride, though they're not able to clear an obstacle that's both high *and* wide.

Weaknesses to Work On

Deer can jump incredibly high, and they can leap long distances. But what deer cannot do is a high jump and a broad jump at the same time. That gives the fence an advantage.

Another factor that limits which fences a deer will test is what is immediately on the other side. Deer will not jump over a fence without a reasonable expectation that a safe landing awaits. Accomplished leapers that they are, they know the inherent damage of a bad landing, and avoid putting themselves at risk.

Elements of a Great Offense

Deer do not understand the difference between a fence and any other obstacle. If something is blocking their way to something desirable, like a garden full of food, they will go over, under, around, or through the blockade every time. If it isn't, they won't. Our job is to teach them that their efforts spent trying to get into the garden are much greater than any reward once they get there.

Fence building is not rocket science, but knowing the right materials, tools, and techniques can make the building experience quicker and less expensive, and the finished product stronger, more durable, more attractive, and more effective. All deer fences have a few common denominators: posts and fencing wire, braces

Setting the Fence Line

The first step to building any fence is laying out the fence line. If your fence will follow your property line, be sure you know the surveyed boundaries of your property. Depending on your relationship with your neighbors, it might be worth a visit to local authorities to verify your property line. Make a point to set the posts just inside the line, so that the width of the posts (from four to eight inches) doesn't accidentally cross the line when the fence is installed.

To install a straight fence line, set or mark the corner posts, then string a line between them. Mark off spots where posts will go at regular intervals, depending on the type of fence you are building. Always check with your local utility and phone companies before digging fence postholes.

to beef up their stability at the corners, and fasteners. The equipment you need to construct a fence will depend on the type of fence you intend to build. A single 30-inch-high electric-wire fence can be built with little more than a hammer (to pound fiberglass posts into the ground) and a pair of wire cutters. However, a massive estate fence, constructed with pretreated wooden posts, may require the services of a hydraulic post-pounder and a tractor to operate it. If putting up high-tensile wire, you'll need one of the many wire tighteners on the market, as well as tensioners to keep the wire taut. The consensus, from commercial growers to backyard gardeners, is that a well-designed, properly constructed fence is your best bet for physically barring deer from your yard.

Fence posts. These may be wooden, metal, wood-plastic composite, fiberglass, or plastic, with the latter two being appropriate for lightweight netting or electric fences as they are not designed to withstand the forces of stretched wire.

Wooden posts can be purchased in a range of widths and lengths, with eight-foot long, four-inch diameter being the most common. When pressure treated with wood preservatives, these posts will last years longer than untreated wood. Some

Digging Holes

Any good engineer will tell you that posts driven forcefully into the ground are more secure than those that are set into a hole and then backfilled. Considering that labor is the most expensive part of any fence, post-pounders — though they average around $10,000 to buy and require the use of a tractor to operate — may be a great investment when you weigh the expense against the time it takes to dig or even auger out holes. In many areas, post-pounders are available from equipment-rental businesses, as are the tractors needed to operate them. They work by ramming the posts, wooden (with one end honed to a point) or metal, into the earth by sheer force. Many are designed for "one-man" operations.

The alternative for wooden posts is to dig a hole at least two feet deep for each post, place the post in the hole, then backfill the dirt while checking periodically with a level to be sure the post is straight, all the while tamping down the dirt to make as tight a fit as possible. Another option is to dig the hole, then backfill with cement, which provides a sturdier post than one backfilled with soil.

Steel T-posts are designed to be pounded into the ground with a handheld steel-post driver that a single person can operate. Just be sure to wear gloves and earplugs — the sound of metal ringing against metal can give you a headache! Steel posts can even be driven in with a sledgehammer, but that takes more strength and better aim.

alternatives to pressure-treated wood are hardwoods like red cedar, white cedar, black locust (which can last for years without treatment), and posts made from wood-plastic composite material. You can also use linseed oil to protect wood posts from rot. Boil enough linseed oil to cover the number of posts you'll be using, then stir in pulverized charcoal until the mixture takes on the consistency of house paint. Coat the posts, let dry, and use as you would commercially treated posts.

Some wooden posts are sharpened at one end like a pencil tip. If you set them in the ground with the sharpened end up, they resist rotting from settling snow or rain; otherwise, pound them into the ground for a more secure fit. Used railroad ties, if in good condition, make sturdy though rustic fence posts and are prized as corner posts. Steel T-posts are designed for livestock fencing, with bumps down one side to help secure clips used to hold the wire in place. Rebar and angle iron can even be used for net or electric fences. Galvanized pipe posts come in a variety of sizes as well, and last 40 years — longer than any deer!

Braces. The weakest points of any fence are the corners. Therefore, any fence design needs to use beefed-up materials to strengthen, or brace, these areas. This is especially true of fences on which wire is stretched tight, because the force of the taut wire pulls in both directions against the corner posts.

Corner posts should be no less than five inches in diameter, pressure treated to withstand the elements, and sunk deeper than those in the fence line — at least 42 inches. A post set 42 inches deep is twice as secure as one set 30 inches. Set end and corner posts holding high-tensile wire 48 inches deep. Several brace configurations are possible, including a double horizontal brace assembly that will stand up to high-tension wire. Double braces

Topping It Off

It's often said that the higher the fence, the more effective it will be in keeping out deer. But it's not quite that simple. One factor affecting the height for deer fencing is visibility, especially the top of the fence. Deer have poor depth perception and if the top of a fence is difficult for them to see, they most often will not challenge it. For this reason, it is not advised to put a solid rail or strongly visible wire along the top of a fence horizontally. It is recommended, however, to mark wires with ribbons to show that a fence is there, but not to give away the maximum height. In fact, one researcher suggested planting "false limits" by tying ribbons above the top of the fence on wire extensions. A deer's eyesight is one of the few weaknesses you can take advantage of.

Six Quick Fencing Lessons

Lesson One: When possible, fence first, plant second.
Lesson Two: If deer can't go over, through, or under, they'll go around.
Lesson Three: A fence with a hole is no fence at all.
Lesson Four: Fences over bumpy or steep ground are more difficult to install and can be more expensive in terms of both materials and labor.
Lesson Five: Pounding beats digging. Posts driven or pounded into the ground are more secure than those placed in a hole dug out by hand.
Lesson Six: There's more than one way to fence out a deer!

will withstand more force than single ones. Cross members should be no longer than eight feet; the shorter the brace, the more force it can withstand.

Fasteners. The type of fastener you use will depend on the type of posts and fencing you choose. The old standard for attaching wire to wooden posts is the steel staple, invented sometime prior to 1873 (when a patent for barbed wire was finally issued to Joseph F. Glidden, of DeKalb, Illinois). Barbed staples hold better than smooth, and galvanized last longest in soft woods, such as pine. Fence clips are designed and marketed specifically for attaching wire to pipe or steel posts. Special fence systems, such as New Zealand fence, deer netting, and vinyl fencing are sold with their own fasteners.

Nonelectric Fences

The following nonelectric fences, tested in the field by researchers and gardeners, have proven effective under the conditions noted for each. The intention here is not to provide you with all the information you need to actually build a fence — for that, consult a fencing book, such as *The Fence Bible* by Jeff Beneke (Storey Publishing, 2005), and local fence builders. But you should come away

Down on the Farm

What better way to learn about fencing deer out than by looking at the ways deer are fenced in? People who raise deer commercially have high stakes for keeping deer (which includes species from blacktails and whitetails to elk and reindeer) on their own side of the fence. The Ministry of Agriculture and Food in Ontario, Canada, covers deer farm fences in detail in a fact sheet. To keep deer in, they recommend the following:

Deer Species	Fencing	Posts
White-tailed deer	8-ft. tall w/ 20 horizontal lines of wire	6-in. diameter, 12-ft.-tall posts sunk 3 ft. deep
Elk	6 ft. 3 in. tall w/17 horizontal strands; two additional strands of high-tensile wire strung at 6 in. and 12 in. above the main fence (for a total height of 7 ft. 3 in.)	same
Black-tailed deer (specifically, the northern Sitka variety)	6 ft. 3 in. tall w/17 horizontal strands and one additional horizontal high tensile strand	same

In all cases, 3.5-inch galvanized, wire mesh with knotted joints between the vertical and horizontal wires is preferred for strength and durability. The mesh should graduate in size, with the smallest openings towards the bottom.

from this discussion and the subsequent section on electric fences with a clearer idea of what will meet your needs.

Standard Deer Fence

The standard deer fence is one that is tall, sturdy, constructed of woven wire, and built at least eight feet tall. The wire can be hog wire, field fencing, or sheep wire with openings up to six inches square, or (more expensive) horse wire, in which the openings are about two by four inches. The heavier the gauge of the wire, the more likely it is to withstand attempts by livestock to breech it. The lower the gauge number, the heavier, sturdier, and more expensive the wire. Some gardeners resort to chicken wire (hexagonal woven wire), with openings of one or two inches (17 or 18 gauge). The smaller weave and heavier gauge is stronger. Woven wire is stronger than welded wire and more often recommended for fencing in livestock, which tend to lean on or rub against fences. Welded wire may suffice for deer.

Wire comes in large, heavy rolls in varying lengths and widths. Hog wire typically comes in four-foot widths, making two widths necessary to construct an effective deer fence. A second width is attached atop the first to be sure it is as strong as the fence and just as long lasting. A row or two of single-strand wire is often attached above that. If possible, the bottom width should be partially buried or, at the very least, pegged down at intervals of six feet to prevent deer from squeezing under.

Because posts must be buried three to four feet deep to stabilize this fence, it requires 12-foot-tall posts and lots of digging or the use of a hydraulic-powered post driver. Be sure to purchase pressure-treated posts for the greatest longevity.

Pros and cons: This type of fence, the most expensive to build and the most long-lasting, is very effective even where deer pres-

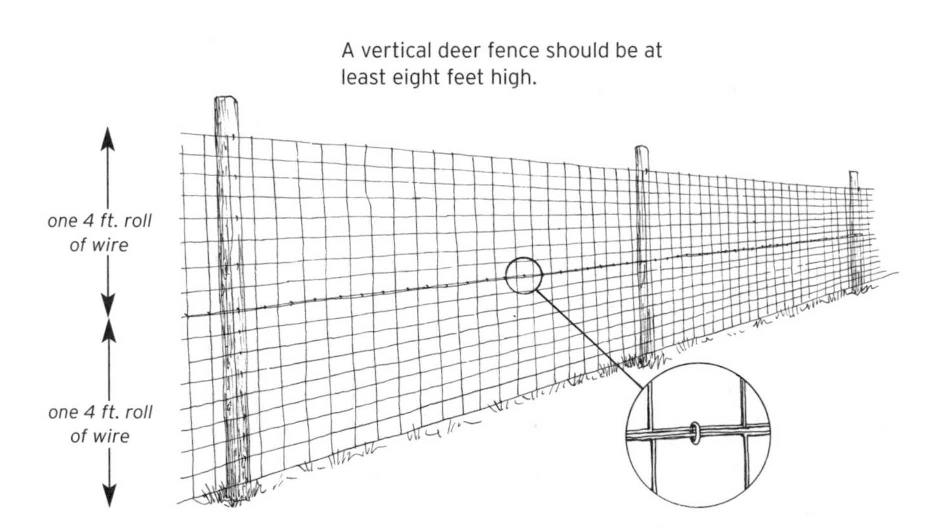

A vertical deer fence should be at least eight feet high.

sure is high. But such fences make their own landscape statement and may be too conspicuous for many homeowners.

Slanted Deer Fence

The slanted style of fence takes advantage of the fact that although deer can clear the height *or* the width of the fence, they cannot clear *both*. The design of this fence encourages deer to walk underneath the overhang, and often they don't even realize that there is a way over the fence. A single width of four-foot wire fencing, supplemented by several rows of single strands across the upper reaches of the overhang, is usually adequate. The fence should meet the ground at an angle of about 45 degrees and reach a total height of six feet with a spread of four feet.

Recommended for areas with from moderate to high deer pressure, a slanted fence protects large areas well. You can make this fence even better with the addition of high-tensile wire.

Pros and cons: The cost of construction is still substantial, but so is the life expectancy of the finished fence. Such fences need not be nearly as high as standard deer fences; however, they take

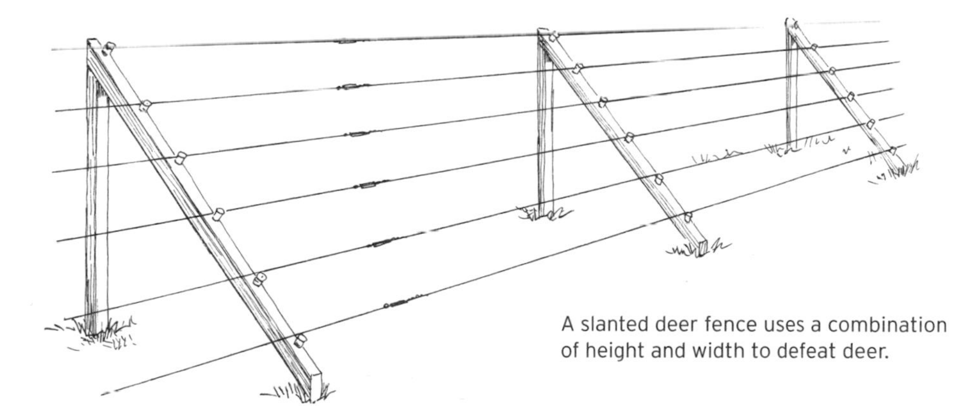

A slanted deer fence uses a combination of height and width to defeat deer.

up about six feet of horizontal space, which in some settings is unacceptable.

Double-Row Fence

To take advantage of a deer's inability to clear the high, wide, and mighty, construct two smaller fences side by side. This design also plays on the deer's fear of leaping without landing. The fences need be only four to five feet tall, and should be placed four to five feet apart.

Pros and cons: Why choose two when one will do? Because unlike a slanted fence, a double-row fence doesn't waste space. You can landscape or garden the median because deer don't attempt to clear either fence.

Those who have tried them give double fences rave reviews. The lower fence height makes it easier to construct than the standard eight-foot deer fence, but it does require more labor and twice the number of posts.

A double-row fence consists of side-by-side vertical fences.

Converting an Upright Fence

If you happen to already have an upright fence in place but it isn't quite tall enough to prevent deer from jumping over, convert it to a deerproof fence. One option is to just keep going up, extending the posts with lashed or bolted poles or boards and attaching another width of woven wire. If you choose to use multiple rows of single-strand wire instead of mesh, the rows should be no more than four to six inches apart. Another option is to attach the post extensions at an angle, either horizontally or at 45 degrees. The first option takes advantage of the deer's height limitations, whereas the two angled designs create a combined high and broad jump. A third alternative is to sink tall poles at intervals along the fence line so that the posts rise eight feet above the ground. In general, this requires 12-foot-tall posts sunk to a depth of four feet. String taut wire at six-inch intervals or use woven wire with six-inch mesh along the tops of the posts at whatever height is needed to compensate for the height of the existing fence.

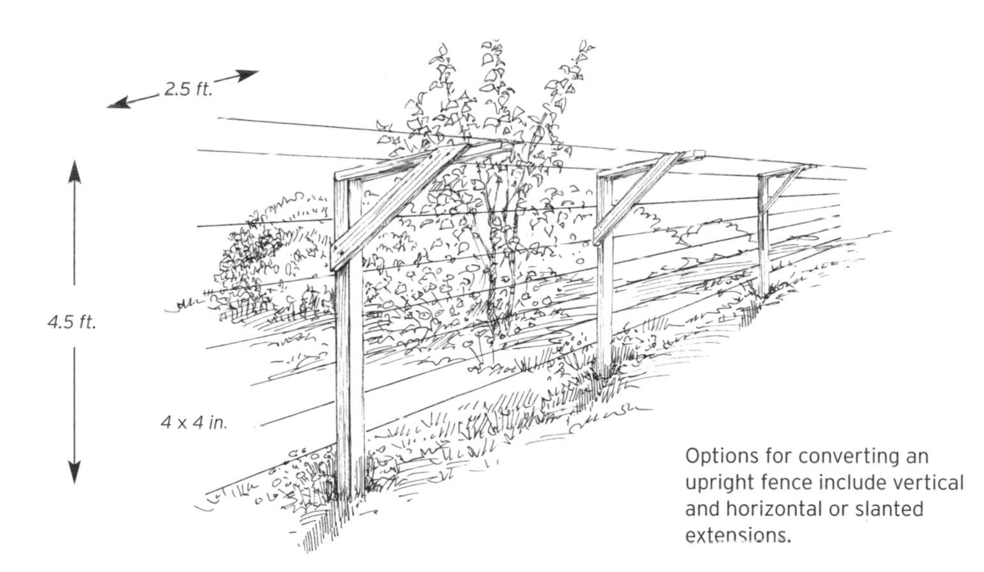

Options for converting an upright fence include vertical and horizontal or slanted extensions.

Pros and cons: Converting an existing fence eliminates wasting what is already there, which cuts the costs of construction considerably. Wood fences are easier to alter than steel, as steel must be bolted or welded. The original fence must be solid and stable. Don't try to extend a fence that has already outlived its usefulness.

Opaque Deer Fence

When is a five-foot-tall, single-width fence sufficient to keep deer on their own side? When they can't see over, through, or around it. Privacy fences look great and can take advantage of this simple principle. Just be sure that the fence has no gap, such as a loose gate or missing boards, which may invite the deer in.

Pros and cons: Solid fences can be constructed of various materials, from attractive cedar or redwood, to rough wood slabs, to leftover tags of sheet metal or fiberglass. The costs vary accordingly. Vinyl fencing, which never needs painting, comes in many prefab designs that make wonderful, attractive, long-lived garden fences, but at a premium price. Wooden fences offer the added benefit of attractive privacy, but if you define a beautiful fence as one that keeps out deer, any material will do. Be aware that using galvanized nails with cedar or redwood will cause discoloration and breakdown of the fence: Over time, the oils in these woods

A solid wooden fence can be both attractive and deerproof.

Prefab deer netting may be the easiest solution to your deer problem.

interact with the zinc of the galvanized metal, causing it to discolor and deteriorate. Paint a fence of salvage material to make it more appealing, or camouflage it by planting climbing vines that you know deer won't eat.

Prefab Deer Netting

The gardening market has offered plastic mesh made especially for deer control for several years now. Proponents claim deer netting has many advantages over bird netting and standard fencing materials. Made from black plastic, the mesh visually blends into a varied background well, making it especially suitable for landscape protection. Some brands are of heavier grade material and are sturdier than others (some will even withstand direct impact or attempts to climb over or through provided the supporting structures are sufficiently rigid), but overall deer netting is most often recommended for areas of low to moderate deer pressure. According to Canadian researchers for airport fence security, the most promising type of plastic fencing is an extruded polypropylene mesh manufactured in Italy by Tenax. (Marketed by Fickle

Hill Fence and Supply in Arcata, California, and Benner's Gardens, in New Hope, Pennsylvania.)

Tenax C-Flex fencing is sold in standard (2¼- × 2¾-inch mesh) and heavy-duty (1½-inch-square mesh) varieties, both 7½ feet tall. The heavy-duty type was designed and developed to resist tearing even in a full-out crash from a running deer. Because the mesh is so much smaller, the heavy-duty product is reportedly 65 percent stronger than the standard, which is important in situations in which deer physically test the fence. Because plastic is notorious for getting stiff and breaking down in sunlight, the polypropylene is specially treated for resistance to UV light. Properly installed and maintained, this type of fence should last about 10 years. The cost can be relatively high.

The fence works on the basis of height and poor visibility. The deer can't be sure how tall it is in order to estimate a take-off and landing. From their vantage point, three to four feet off the ground, it appears very tall but blurs visually the higher it goes. Some deer may test the fence precisely because they can't clearly see the upper reaches of it; others will find that too discomfiting to challenge.

The plastic mesh was designed to be installed in several different ways, making it convenient for an array of settings. Stretch it from tree to tree or between treated or steel posts. Because it is so much lighter than wire mesh, you can place supports farther apart, generally 10 to 15 feet. Also, because the netting is much more flexible than wire mesh, it covers uneven ground better, dipping into hollows and following along ridges and bumps. Special self-locking ties are sold to attach the mesh to supports, and you can buy specially made pegs for anchoring the netting to the ground, to prevent deer from squirming underneath (but in a pinch, bent coat hangers will do the trick).

The netting should be tightened only enough to prevent it from sagging, requiring much less strength in the supports used to hold it up. In cases where deer are likely to challenge the fence by running into it or trying to climb over, this is counterproductive, but in areas of light to moderate pressure, the netting has performed well when mounted on everything from trees to rebar posts. The trade-off is that the lighter the supports, the less visible the fence will be, but also the less of a physical barrier. Benner's recommends running a nylon cable through the top of the netting to help keep it taut, and at about three feet high (the height at which deer are most likely to come into contact with the fence) to help secure the netting. Tighten the nylon and fasten with cable clamps.

Instructions for installing the fence recommend attaching white flags to it at a height of four feet, at 12-foot intervals. This warns the deer that an otherwise invisible barrier has been installed and takes advantage of the "white-flag" signal that white-tailed deer use to warn each other of impending danger. After a month or two, remove the flags, as the deer will have become familiar with the fence and altered their route accordingly.

The netting should be inspected on a regular basis — once a week is not too frequent during peak deer activity in spring and fall. Repair tears and holes in the fence with the ties sold to fasten it to supports or with patches of netting held in place with the ties.

Pros and cons: Lightweight, less expensive, and easier to install than livestock fencing, the plastic mesh blends into the landscape. Although strong enough to withstand deer pushing against it, the standard mesh will give if it is hit at an all-out run. It has not been used in game farms, highway, or airport situations, so heavy-duty dependability is not ensured. No independent studies have been done to test the netting or to determine its vulnerability to chewing rodents or other animals. Until such testing has been done, experts are cautious in recommending it. Even so, Benner's has never had a reported case of the heavy-duty variety of netting breaking since the company started selling it in 1996. In areas where other animals or deer are known to challenge a barrier, wire mesh, even poultry fencing, is preferred. But for the average garden situation, where maintaining the visual integrity of the landscape is important, this is a great solution.

Electric Fences

Electric fences deserve a few words of both praise and caution. Although useful and effective at deterring deer, livestock, and other pests, they carry with them a margin of liability. Some municipalities frown on them, even though the fence chargers built today utilize low impedance and are capable of delivering only a harmless jolt, unlike the shocking monsters of the past. Some areas outlaw them outright. Know your local ordinances before erecting any type of electric fence.

Electric fences offer three advantages over nonelectric types:

- They are inexpensive. The labor and materials to construct an effective electric fence are generally much less than to build an equally effective nonelectric fence.
- They "teach" deer to avoid the area. One good zap and deer head for friendlier territory.
- They can be made either temporary or permanent, depending on the situation.

Electric Fence Components

Specific materials and quantities needed for an electric fence vary with the size, length, and style of fence you erect, but many of the basic components remain essentially the same. Electric fencing supplies, including a fence charger to electrify the wire, a grounding rod, wire, posts and clamps or insulators to attach the wire to the posts, are readily available at most feed, tack, and hardware stores, and even over the Internet. A voltage meter, or electric-fence tester, is a good investment. Use it to check the fence line for charge by touching it to the wire and reading the meter. Digital models give an exact reading of the voltage output for a slightly higher price than traditional models that aren't as precise. Another good addition to any electric fence is a lightning diverter. Depending on the length of the fence, several diverters should be placed at intervals along the fence line to protect the charger from a power surge in the event of a lightning strike anywhere along the fence.

Fence chargers. There are three types of fence chargers. 1) AC chargers that are plugged into a 110-volt electrical outlet (so remember to install them within reach of an outlet). They must be kept dry and are often installed inside a garage or other protected area. They can deliver anywhere from a barely noticeable twinge to a substantial jolt, and can run constantly without the need for recharging. 2) Battery-operated chargers attach to a 6- or 12-volt battery. They have the advantage over 110-volt chargers in that they don't need to be located near an outlet, though some models also require protection from the elements. Twelve-volt batteries work best for fence chargers and can usually be left at least three weeks before they'll need a recharge. Car batteries are sometimes used for this purpose, but are designed to be continuously recharged and can be damaged if discharged too fully. Deep-cycle, marine batteries are a better choice, as they can be almost fully discharged on a regular basis without weakening. 3) The third option is a solar-rechargeable battery, which can theoretically run and recharge indefinitely, but often carry only a 90-day to one-year warranty. The more powerful the unit — and thus the greater the shock it delivers over the longest distance of fence line — the more expensive.

How much shock a charger puts out depends on the voltage and the amperage of the unit. Voltage is a measure of the force with which the current flows through the wire. The higher the voltage, the farther it can flow before the resistance of the wire slows it down. Amperage is the measure of how strong the current is. The higher the amperage, the more uncomfortable the shock. A good analogy is that of water flowing through a hose. Substitute the water pressure for the voltage and the amount of water traveling through the hose for the amperage. Water rushing through a one-inch-thick hose at 100 miles per hour can knock you off your feet; the same amount of water pushed through a ¼-inch-thick hose would certainly smart, but hardly cause injury. Similarly, a large charge traveling slowly (like a large volume of water going through a fat hose) produces only a mild shock. For instance, the static shock you receive from walking across carpet and touching a doorknob is not enough to do you any harm, even though it can deliver 5,000 volts of electricity, because that's at very low amperage. Conversely, the electric chair delivers only about 2,000 volts but at an extremely high amperage. Because the goal of electric fences is to teach the animal, not fricassee it, they are designed to operate at fairly low amperage and higher voltage. To deter deer, a minimum of 4,000 volts is recommended.

A charger emits a brief pulse of electricity, roughly every second, through its terminal, which is connected to the fence wire. This is followed by a very brief period (about a second) when no electricity flows through the wire. The intensity of the pulse varies with the model of the charger. Those with a high-voltage, short-duration pulse are the best choice for deer fencing, as they are harder to ground out by vegetation than lower-voltage, longer duration types. They are also considered safest, because even though they can deliver a shock upward of 4,000 volts, the duration is so short (about two-ten thousandths of a second) that they are least likely to start a fire.

Though electric fence chargers vary in cost and the length of fence they can electrify, all are regulated as to how much shock they can deliver. "Hot" fence chargers, which used to be common, are no longer sold because they present a fire threat.

Grounding rods. These are the Achilles' heel of any electric fence. Without a good ground, the fence will not be charged, at least not well, and your charger will eventually burn out. Commercial grounding rods are typically solid metal, ⅜ inch to ⅝ inch in diameter and six to eight feet long. Copper pipe works well, too, but it's expensive, and often a copper-coated steel pipe is substituted. Don't use a painted post (such as a painted metal fence post), as the paint acts as an insulator and does not conduct electricity. Utility ground or water pipe is not recommended either. However, under drought conditions, when there is not enough moisture present in the soil to conduct electricity, pipe makes a good choice because water can be poured directly down the pipe to improve the grounding system.

Wire. The most common sizes of wire used for electric fencing are 14- to 17-gauge wire. The heavier wire (with the lower gauge number) costs more, but is sturdier and carries the current farther and faster because of lower resistance. The types most commonly used are galvanized and aluminum. Galvanized wire is steel wire that has been coated with zinc to prevent rusting. Not all galvanized wires are created equal. The wire with the heaviest coating of zinc is categorized as class III. It lasts the longest and generally costs the most. Class I galvanized wire has the thinnest coating of zinc. Aluminum wire is generally the most expensive, and has the advantages that it won't rust, weighs about a third of what steel wire does, and conducts electricity four times faster. The disadvantage is that it breaks easily.

Posts. Just about anything that can hold up wire can be used to build electric fences, but certain types of posts have traditionally been preferred. Wooden and metal posts require special insulators to connect the wire. Special plastic or fiberglass posts are marketed for electric fence construction and since they don't conduct electricity, the wire can be attached directly to the posts.

Insulators. Nowadays made of plastic, insulators are nonconductive attachments used to connect the wire to fence posts and to strategically place the wire at different lengths away from the posts. Without insulators, the wire would ground out on wooden or metal posts. Some insulators are only an inch or so long; others are six inches long, and are designed to hold the wire away from the post. Insulators made for wooden posts should be nailed onto the post. Those made for steel T-posts are clipped to the ridged side of the posts. In the past, insulators were made of glass or ceramic, but plastic is cheaper, lighter, and easier to ship and store without breakage. You can even fashion your own by recycling rubber garden hose. Cut it into sections and poke a hole through the center through which to thread the fence wire.

Electrifying the Fence

Constructing an electric fence begins with understanding how the charger works and installing it properly.

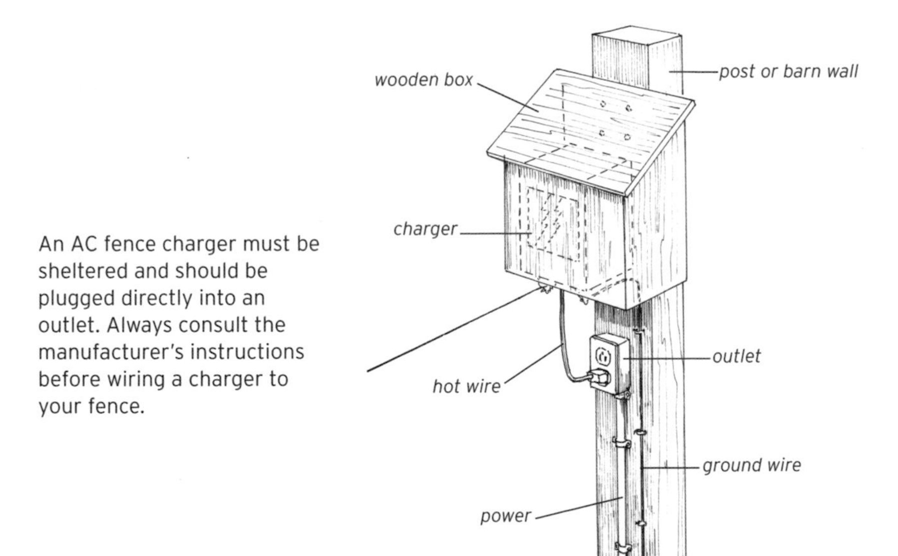

An AC fence charger must be sheltered and should be plugged directly into an outlet. Always consult the manufacturer's instructions before wiring a charger to your fence.

An AC fence charger must be protected from the elements and installed where there is no danger of mechanical damage or of starting a fire, and, of course, where children and animals can't reach it. Another consideration is positioning the charger near an electrical outlet. It's best to plug the charger directly into the outlet, rather than to run an extension cord between charger and power source.

To work, all electric fences must be grounded. For the grounding system to achieve optimum performance, manufacturers typically recommend driving at least three grounding rods (more if the fence is long), each about six feet long, into the ground about ten feet apart. The rods are then connected to one another and the fence by a continuous wire. If the ground is rocky or too hard to drive a rod six feet straight down, experts recommend driving several rods into the ground at a 45-degree angle in a wheel-spoke pattern. Be sure the ground system is installed within 20 feet of the charger. Use ground clamps to attach the grounding wire to the ground rod and be sure to check the rods every two years or so. They will rust underground, and a rusty ground rod is useless, so replace as necessary. Also, be sure to keep fence grounding rods at least 50 feet from any other grounding systems, as one will interfere with the other.

Electric fences use one of two basic shocking systems. In the first, an "earth-return" system, the current passes from a live wire, through the animal, and into the ground. From there it travels to the grounding rods and then back to the charger, completing the electrical circuit. This type of system works best in areas where the soil is relatively moist all year.

Dry soil requires a different type of shocking system. Because dry soil disperses current more erratically than does moist soil, it requires a "wire-return" system that doesn't rely on the conductivity of the soil. This system delivers a shock when an animal touches two wires: one live, the other grounded. Some electric-

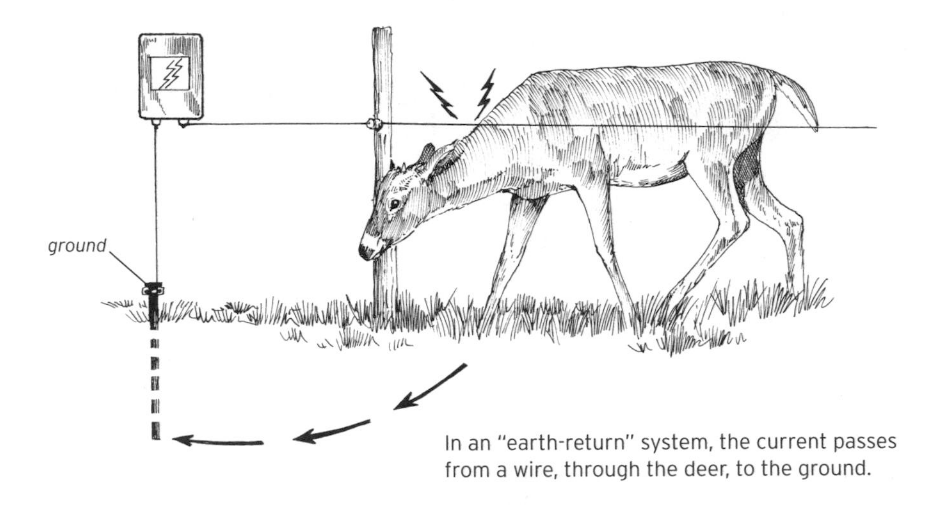

In an "earth-return" system, the current passes from a wire, through the deer, to the ground.

fence designs call for electrified wires placed as high as five feet above the ground. Chances are that if a deer hits the fence at this point, the deer is airborne (not grounded) — another case when a wire-return system is the best bet.

Installation and Operation

Before constructing any electric fence, be sure the path of the fence is clear of tall grass, weeds, brush, and other debris that might accidentally come into contact with the electrified wires. String a strand of twine or wire between two end posts to serve as a guide to ensure a straight fence line. Set or drive fence posts about 15 to 20 feet apart, depending on the terrain. Once the posts are in place, attach insulators (if needed for the type of post used) to each at the desired height for each wire. If connecting the fence charger to the fence wire requires running a wire underground, be sure to use standard insulated wire (such as Romex) and to thread it through a PVC conduit prior to placing it underground. Caulking the ends of the conduit will prevent water from collecting inside the pipe and possibly shorting out the wire.

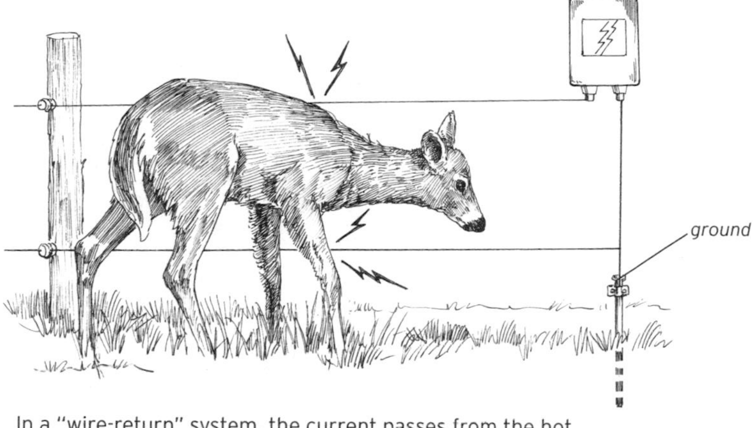

In a "wire-return" system, the current passes from the hot wire, through the deer, to the return wire.

Especially at first, leave the fence on 24 hours a day. Deer frequently travel and feed at night and in the wee hours of the morning. If the fence is turned off, they quickly realize it can't harm them and will walk right through the wires. On average it costs only about $1 a day to operate an AC-powered fence charger, nothing to run a solar-powered, one and a marginal amount to recharge a battery-operated model.

Aluminum wire, though highly conductive, may not be a good choice if your fence lies in an area where tree limbs could fall on it or livestock could run into it. It breaks too easily. A heavy-gauge, electric-fence wire will save many headaches. Wire must never cross with other wires or touch fence posts, trees, or other objects. Also, be careful not to install the wire within half an inch of any metal surface, including steel fence posts, gate hinges, or nails or fence staples in fence posts. The current can arc though the metal and ground out the system at that point, a tricky failure to find. Also, thoroughly flag the fence line; colored surveyor's tape, tied around the wire at intervals, works great. It's cheap and very visible. Not only will this tip off the deer that a new barrier is in place and speed their recognition of the boundaries, but it will also remind the neighbors where your fence is.

Electric fences require careful routine checks to be sure nothing has come into contact with the wire. Something as seemingly insignificant as a blade of tall grass can complete the circuit and ground out the system, thus rendering it useless. Be especially vigilant in checking your fence lines after a storm, as grass or brush can blow onto the wires and cause a short. Always turn off the charger before attempting any repairs to the fence. And don't forget the simple things, like checking to be sure the charger (if an AC model) is plugged in or that corrosion has not formed around the battery terminals on battery-operated models.

Making a Good Thing Better

There are several ways to improve the effectiveness of an electric fence, all of which expedite the learning experience for deer. One way is to be sure the fence is well marked, so that deer can see it and quickly learn which specific area is off-limits. Electrified ribbon (polytape) will mark the presence of the wire. Polytape is more visible at night than surveyor's tape and will zap any deer that come into contact with it. Another helpful tutoring aid is to lure the deer close enough to the fence to get a shock. Smearing a mixture of peanut butter and oil on polytape attached to the hot wire both lures and reprimands approaching deer. One fence model was devised to bait a deer into sniffing or licking the fence and receiving a substantial shock on its tender wet nose or mouth. Who would stick around for more of that?

The reverse tactic has also worked well to improve electric fences. Instead of drawing deer near the fence, Dr. Milo Richmond, with the New York Cooperative Fish and Wildlife Research Unit, experimented with commercial deer repellents in combination with electric fences and found they significantly increased the effectiveness of the fences. He tied cloth strips sprayed with odor-based repellents every three to four feet along the fence line and compared the results to fences that were baited with a peanut butter lure. The stinky fences actually proved more effective in these trials than the baited fences, as they constitute a double negative-conditioning regime of an offensive odor combined with an aversive electric shock. He also found it less of a mess and bother to work with commercial deer repellents than to mix and smear peanut butter and oil. Don't forget that repellents need to be reapplied about once a month. And, of course, any fence does better at excluding deer if it's put up before the goodies are planted inside it.

The Electric Strand

Going back to the premise that deer find it very disconcerting to walk into something unseen, they find it all the more discomforting if the "something" carries the sharp, sudden pang of an electric shock. In many garden scenarios, a single strand of unseen electrically charged wire can send deer darting off.

To use: String fence wire to posts positioned around the perimeter of the yard at a height of about 30 inches and attach one end to a grounding rod. Wire the fence to an electric fence charger. Be sure to follow the manufacturer's instructions and to keep the charger (unless it is a weatherproof solar model) under cover from the elements. When a deer touches the wire, it completes the electrical circuit and suffers the consequences.

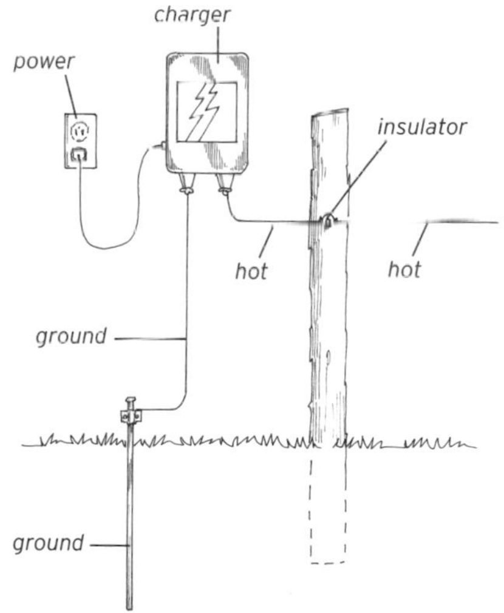

Key components of an electric fence are posts, wire, insulators, a grounding rod, and a charger.

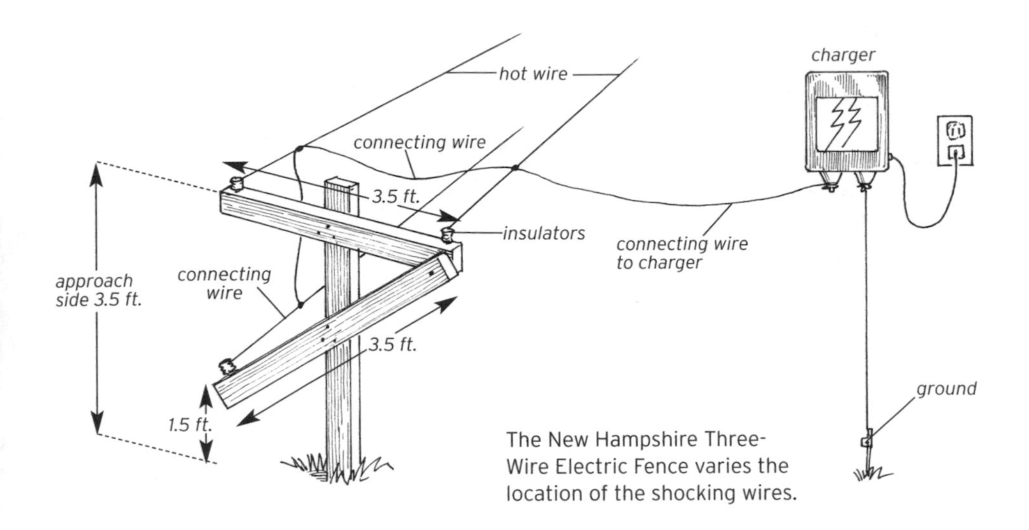

The New Hampshire Three-Wire Electric Fence varies the location of the shocking wires.

Pros and cons: An electric strand is lightweight, flexible, easy to install, nearly invisible when in place, and surprisingly effective at deterring unsuspecting deer. The greatest expense is for the electric charger. Some deer, however, especially in times or areas of extreme deer pressure, will brave the shock or learn to clear the strand. Periodically altering the location or height of the wire helps tremendously.

Snow cover, which can insulate the deer from the ground, may disable an electric strand even if the wire itself is above the snow.

Notify neighbors, especially those with small children, that your property is guarded by electrified wire. Consider turning on the charger only from dusk till morning, when deer activity is greatest and child activity lowest.

New Hampshire Three-Wire Electric Fence

The New Hampshire Three-Wire Electric Fence utilizes the high-and-wide approach, and is recommended for areas with low deer pressure. The angles of the design place the wires at different levels and widths. As illustrated, the fence is made from one-by-two-inch wood (scraps or new lumber), nails, insulators, wire (preferably 18-gauge copper-covered steel-core wire), a charger, ground wire, and grounding rods. This design is best in areas that receive little or no snow, as it is close enough to the ground to be shorted out by heavy snowfall.

Another version of the same basic design uses alternating rows of fence posts but attaches the wire according to the same dimensions shown.

Penn State Five-Wire Outrigger Electric Fence

The Penn State Five-Wire Outrigger Electric Fence design is recommended for small-to-moderate acreage with moderate deer pressure. It calls for high-tensile New Zealand wire (12½ gauge, 200,000 psi) and special in-line wire tighteners that maintain the tension at 250 pounds. Special crimping sleeves are used wherever the wire is spliced to maintain the strength of the fence. Sturdy corner posts and supports are necessary to withstand the high tension, which ensures that the wire can absorb the force of deer running into it. Precise spacing of the wire prevents deer from going through the fence, while the height and width, as determined by electrified outriggers, ensure that deer won't go over. The lowest wire on the fence is set 10 to 12 inches from the ground,

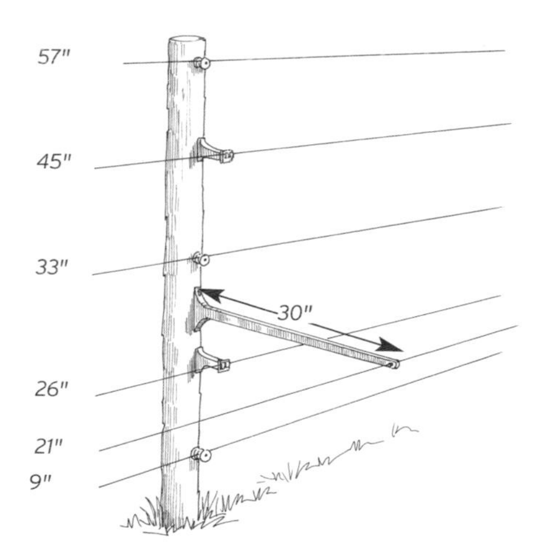

The Penn State Five-Wire Outrigger Electric Fence uses high-tension wire as well as electricity to discourage deer.

which means all brush and grass must be cleared from the path of the fence and kept down. A high-voltage, low-impedance charger grounded with about 20 feet of rod or pipe gives the system its shocking power.

Variations of the fence have been used in different settings — for example, shorter versions serve well to protect small areas. If necessary, standard electric-fence wire can be substituted for the New Zealand wire. The fence won't be as strong but it will be less expensive to construct.

Minnesota DNR Electric Fence

A simple electric fence developed by the Minnesota Department of Natural Resources proves that we can outwit deer. Folks that use it are delighted with the concept, and many have come up with embellishments. Basically, the idea is to lure the deer to the fence and then teach them to avoid contact.

This design works best for areas of under 12 acres that suffer from only light deer pressure. If erected early, before deer pressure becomes more intense, this fence continues to protect your yard because the deer have already learned that it's off-limits.

Before setting up the fence, remove any grass or underbrush that could possibly come in contact with the wire. Set posts

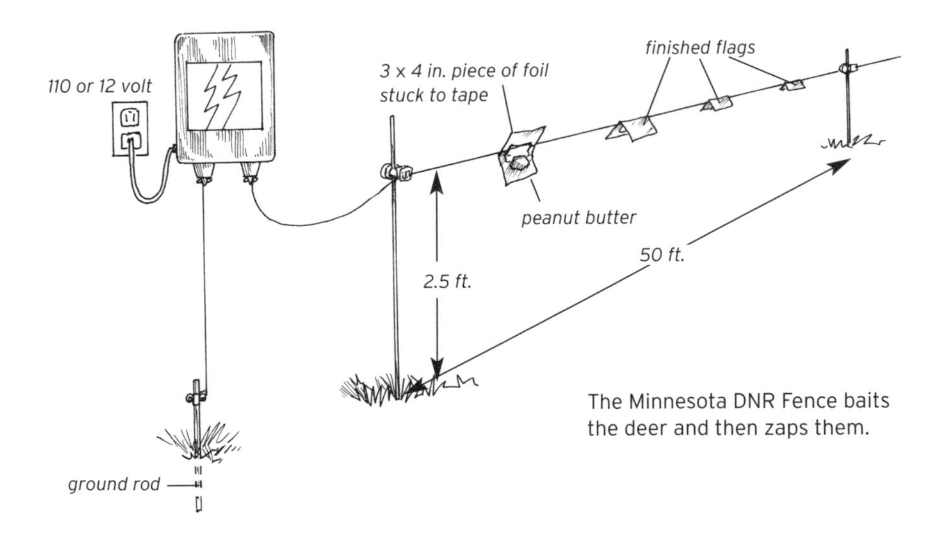

The Minnesota DNR Fence baits the deer and then zaps them.

When a Fence Isn't

It's a fairly common mistake to fence the yard incompletely. You could call this the open-door policy of fencing; build the fence, but leave the gate open. Many people don't close off the driveway or choose not to spoil a view, or for some reason leave a gap in an otherwise flawless fence. Incomplete fences actually work for a while with white-tailed deer, which tend to stick to their established pathways. If these paths are barricaded, it may take the deer a few days to formulate an alternate plan. But a few days is all it will take. If deer have made your yard a regular stop, an incomplete closure is an inadequate solution. Sooner rather than later they'll find their way through or around such a fence and resume munching the marigolds.

What constitutes an incomplete fence? A 12-inch gap in a fence is room enough for determined deer to squeeze through. If deer can't find a way around or through, they'll look for a way to get under. If you haven't partially buried or pegged your fence to the ground, deer can wriggle beneath. A gap in a gate or open driveway will also render incomplete an otherwise useful fence.

If you combine all these considerations, the logical conclusion is that an effective fence, besides being installed before deer damage starts, must be high, a combination of high and wide, or solid enough to obscure potential landing sites. And, of course, it must be impenetrable.

(wooden or steel) no more than 50 feet apart. The fence uses only a single electrified wire, which hangs about two and a half feet above the ground (if it sags, space the posts closer together). Attach insulators to the posts, string the wire through the insulators, and connect a grounding rod and fence charger.

Now for the secret weapon: the fence beckons unsuspecting deer to get zapped on the nose by baiting them with the irresistible aroma of peanut butter. One sniff and *zzzaappp.* Off they go before they know what got them.

There are several ways to attach the peanut-butter bait. The original idea was to wrap pieces of adhesive tape to the wire at three-foot intervals and attach pieces of heavy-duty aluminum foil (about four inches square) to the tape. The inside of the foil flags were then spread with a mixture of peanut oil and peanut butter to tantalize the deer. As more people have experimented with the design, different approaches have sprung up, including the use of spring-loaded clips to attach the peanut-butter flags. These are quicker to install and less likely to be blown against the post by the wind, which causes the fence to short out. However you attach the peanut-butter bait, be sure to change it every three or four weeks, as it eventually becomes rancid.

Gates and Bypasses

Another way in which we sometimes unwittingly negate a good fence is with a bad gate. Gates must be as impenetrable as the fence or deer will quickly find the opening and invite themselves in.

Gates for wire or mesh fences can be constructed of the same wire or mesh strung across a frame built to the same height as the fence. Attach the gate on heavy-duty hinges to a solid, braced post. Wide gates, such as those for driveways, may need to be supported from above to prevent sagging. A guy wire attached to a higher support post will help to disperse the pull of the gate on the support. Driveway gates can be constructed as double or single panels. Double panels must meet closely in the middle and be secured together to prevent deer from squeezing through.

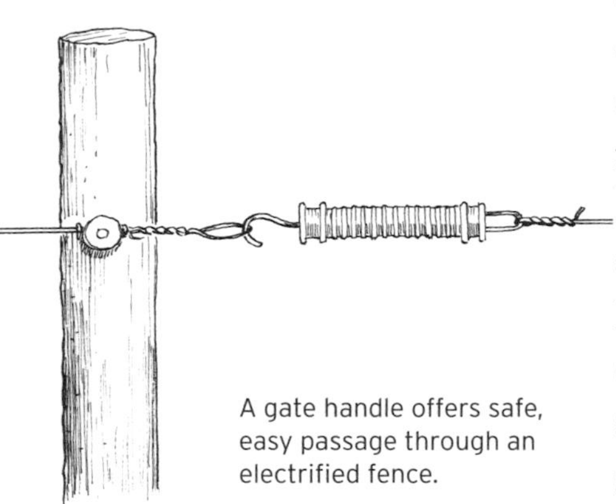

A gate handle offers safe, easy passage through an electrified fence.

Walk-throughs and stiles are options for deerproof fences. Deer cannot maneuver the twists of the walk-through, nor are they apt to march over a stile. When a gate would leave a gap, consider one of these instead.

Electric-fence plans rarely include instructions for gates. Once the current is broken, no more electric fence, so gates must be either electrified or designed so that the current bypasses the gateway. One way to bypass the gate is to run the fence wire high above it; another is to cover the wire with an insulated sleeve (rubber hose works fine) and reattach it to the fence on the opposite side.

Electric gate handles resemble oversized insulators with a hook at one end and an attachment for the fence wire at the other. The handles are spring loaded to keep the fence tension constant and to allow you to open the "gate" easily by pulling up the handle and unhooking it from the fence wire on the opposite side of the gate. Rig the handles so that when you unhook them, the electric current disengages in the handle, the gate wire, and any part of the fence downcurrent from the gate. You can set the handle down without having a live wire on the ground.

Protecting Individual Plants

Fencing in vulnerable or especially valuable plants does not necessarily require fencing the entire estate. As an alternative, protect individual plants with little fences of their own. You can easily fashion wire or mesh cages to protect any size plants — from a row of newly emerging lettuce to the trunks of young fruit trees. Drape plastic netting over susceptible plants or entire beds or borders. But be sure to provide a rigid support. Deer will lean over or snake their heads under or through a loose barrier. Drive stakes on either side of shrubs and wrap chicken wire around and over

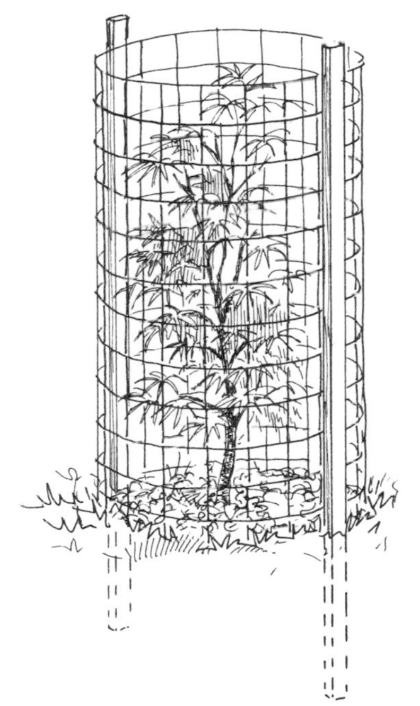

Nylon or wire mesh can effectively protect vunerable plants.

them. Attach the wire or netting to the supports with wire ties for added strength and deer resistance. For portable protection that can be used over and over on different plants, build a framework of scrap lumber or PVC pipe and attach mesh to form a protective cage. Heavier-duty wire mesh, such as hardware cloth (small-mesh wire), can be bent into shape and positioned to protect plants without additional support. Be sure to anchor cages in place by burying the bottom of the framework or weighing down with rocks or bricks so that deer can't nudge them over.

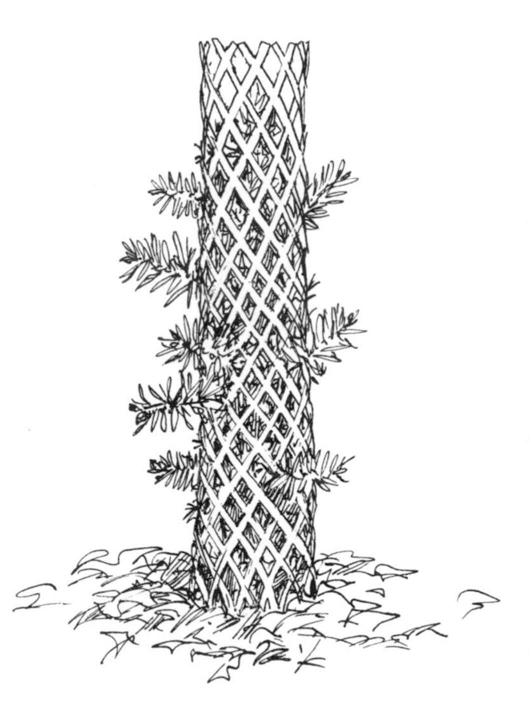

Netted polyethylene tubes, such as Vexar tubes, protect young plants from deer damage.

Plants may not need protection indefinitely. For instance, once fruit has been harvested or small trees have matured, they need not remain covered. Unlike an entire perimeter fence, individual plant protectors are easy to remove.

Protect the trunks of trees and shrubs from rubbing damage caused by amorous bucks by wrapping them with a protective trunk wrap. Garden centers carry products made especially for this purpose, such as Vexar tubes, or substitute strips cut from old inner tubes.

SEVEN

Community Efforts

When it comes to deer, the comfort zones for some gardeners are so very different it's like trying to find reality in the twilight zone. The issues and degrees of concern vary so widely that gardeners in different parts of the country may sound as though they aren't even talking about the same subject when they discuss deer damage. Some gardeners speak of living peacefully with deer that pass through their garden for just a light nibble. For those under truly heavy deer pressure, the problem and their reactions to it are more severe than for those whose gardens haven't been ransacked beyond recognition. In these areas, calls for deadly force ring out. For many, the deer issue goes far beyond any single person or garden to encompass neighborhoods and entire towns — it has become a community concern.

Management Strategies

When the perception of overpopulation is based on the requirements of people, not deer, the solutions are dictated by social concerns, which can be a real pickle for wildlife managers. Rather than natural selection, predation, and the carrying capacity of the habitat, they must contend with annoyed gardeners, frustrated landowners, nervous drivers, and public opinion about how to resolve all types of deer–human conflicts. Citizens with strong opinions rarely agree with each other, let alone authorities, on wildlife management agendas. The good news is that community efforts can have a lot of power in protecting our backyard environments as well as the larger environments and habitats around us. And we can do it all on our own. The front line should always be a coordinated effort by members of the community. This can range from talking with your neighbors and developing a plan for your home and immediate surroundings to attending town meetings. The public can contribute significantly to minimizing deer problems: by utilizing plants that are unappetizing to deer, by protecting individual plants or plots that are particularly susceptible, by not feeding or otherwise enticing deer, and by driving cautiously and defensively.

The issue of deer management is most charged in areas where the deer population (usually white-tailed deer) exceeds the ability of the natural hatibat to sustain it. Intense deer pressure on yards, gardens, orchards, crops, and nurseries becomes a major economic concern as well as an emotional issue. Studies in various states routinely put the estimated crop damage in the tens of millions of dollars — and that doesn't reflect the damage to private yards and gardens. Compound deer overpopulation with a bad winter or decimated forage (which is the predictable result of deer overpopulation), and the results can be gruesome.

Even in areas where no one questions that the deer population exceeds reasonable limits, competing interests have very different ideas about how to lower it. For any successful community deer management plan to succeed, cool heads need to prevail, clear objectives need to be set and data need to be collected to define both the problem and possible solutions. Deer densitity must be accurately estimated. Damage must be quantitified. For instance, if decreasing landscape damage is a goal, specific information is required to determine the current level of damage in order to assess the success or failure of any program that is implemented. Similiarly, if decreasing the number of deer–vehicle collisions is a goal, figures related to that problem must be established.

Next, several factors should be considered in choosing strategies to control deer populations:

- Human safety
- Humane treatment of animals
- Responsible long-term management of deer herds

Human safety. Some methods of deer control can be as dangerous to people as they are to deer. Opening fire in suburban areas is generally not a good choice. However, firearms and other weapons can be a useful and safe tool in the hands of professional sharpshooters and archers whose job it is to eliminate deer in densely populated areas.

Humane treatment of animals. Animal welfare is more than just an issue of conscience; it is supported by federal law. But even killing deer can be done in ways that are more humane than others. One study showed that an accurate gunshot to the head was less stressful than might be assumed. Stress levels in deer killed by a rifle shot were compared with those of deer that were chemically immobilized (method not specified), then euthanized with a bolt gun. By measuring blood levels of cortisol (which increases

Deer Counts Count

Wildlife agencies consider accurate counts of deer critical to management strategies, yet not everyone agrees on the methods of counting or the actual count. Counts may be taken in several ways, such as deer drives, in which deer are herded past posted counters who tally them up, and aerial surveys, which tend to be less accurate.

An inaccurrate count can easily lead to a false assessment of the number of deer that need to be removed. For example, if a certain area can optimally sustain 25 deer per square mile and the count determines there are 240 deer in a four-square-mile area, how many deer should face removal? What if the count was too low? Removing too few deer is a wasted effort: The slight temporary dip in population can trigger more energetic repopulation. What if the count was too high? If only about 150 deer live in the area, then removing 140 of them in an effort to reach that optimal ratio would instead wipe out the population.

under stress) postmortem, it was determined that the deer that were chemically immobilized had levels 10 times higher than those that were shot.

Responsible long-term management of deer herds. Long-term effects on deer populations should also be considered. In cases where animals are selected for elimination, what are the criteria for selection? Hunters may have very different agendas than other members of the community or wildlife managers — most prefer big antlers to scrawny does. If only trophy bucks are removed, the breeding cycles quickly replace the numbers of deer, but the vitality of the herd suffers when prime breeding males are removed from the gene pool. Removing does, on the other hand, although an issue that spurns hot debate, can be more effective in lowering numbers of both males and females by reducing fawning rates. In any deer-management program, cost effectiveness, efficiency, and adherence to existing state and local laws must be factored in.

Nature's Course

The "let-them-be" approach drops the problem back in the lap of nature. Trouble is, people caused this mess and nature isn't all that well equipped to handle such extremes in a kind manner. Under normal conditions, deer populations do not exceed the capacity of the habitat because each spring, before fawns are born, the does drive out last year's babies into the world. This dispersal is one of nature's many ways of keeping the population in balance with the land. But deer "left to nature" in severely overpopulated areas have nowhere to go. Surrounding areas are already packed with other deer. Whitetails, especially, will occupy their ancestral grounds for generations, regardless of the area's condition.

Deer left to nature in such circumstances lead pitiful lives and meet tragic deaths. They gradually weaken, sometimes over generations, until food disappears completely, and then they linger on until death overtakes them. Weakened deer are sickly, often riddled inside and out by parasites, easily devastated by disease, and often in pain from their infirmities. In winter they die by the hordes, breaking their necks on fences that healthy deer haughtily bound over, and falling victim to predators or dogs, sometimes being eaten alive, too weak to resist. Or they simply lie down and do not get up. To suggest that nature will solve this problem is irresponsible, cruel, and ignorant.

One method sometimes suggested for helping nature take its course is restoring natural predators. But the lack of natural habitat for predator species, the likelihood that predators will kill other species, and the fact that predators often leave a designated area for their own reasons all make this a poor solution for all but wilderness areas.

Hunting

To some, hunting is merely a sport, or perhaps a way to put meat on the table. Others consider it a practical management tool. Still others regard legalized deer hunting as unconscionable slaughter. However you look at it, the fact remains that deer hunting has always been a part of the American culture and has evolved into multibillion-dollar industry.

Every five years, the U.S. Fish and Wildlife Service conducts a hunting and fishing survey. Figures from the most recent (2001) survey estimate that more than 10 million Americans (16 years and older) hunted deer and and nearly another million hunted elk. They killed approximately six million white-tailed deer, which hardly dented a population estimated at more than 25 million. Biologists explain that to stabilize a deer population, each year 40 percent of the females must die, naturally or otherwise. To reduce any given population requires that even more deer be taken.

Many areas have tried to stabilize the deer population by regulated hunts — in extreme cases employing more creative approaches. In Wisconsin's "Earn a Buck" program, for example, hunters earn the right to go after a set of antlers by first bringing down a doe. In isolated areas where deer densities have reached intolerable levels, controlled hunts often permit expert hunters to take specified numbers of deer as part of the community effort to reduce the populations as quickly and humanely as possible. In effect, human hunters replace displaced natural predators.

So why doesn't hunting work better as a population management tool? Well, there are several reasons. The country's 11 million deer hunters, of whom half or fewer kill a deer each year, harvest too high a proportion of bucks — and those they take are often the healthiest of the herd, with the most impressive antlers, rather than the weak or inferior. This is the opposite of natural selection, in which the weak, ill, and inferior are the first to go and the strongest are left to perpetuate the population. Hunters also take their prey all at once in any given area, thus creating a short-term dip in the population, which deer instinctively fill. Fewer deer equals less competition equals higher reproduction. This supports the contention of many antihunting groups that hunting management schemes, inadvertently or otherwise, help create overpopulation, thereby maintaining the apparent necessity for more hunting. If human hunters behaved more like their wild counterparts, killing the lesser animals and leaving the fittest to survive, over an extended period rather than just prior to or in the midst of the annual mating season, they might be more successful at affecting the overall numbers.

Another drawback to relying on the human predator for population control is that he can't work in his own backyard. Many of the heaviest concentrations of deer occur in and around suburban housing developments, golf courses, parks, and even college campuses. Hunting in such areas just isn't practical or safe. Furthermore, in many deer-dense areas, the deer are virtually tame, and public sentiment tends to run deeply against shooting them.

One alternative to public hunts is hired guns. Sharpshooters, armed with high-powered rifles or bows and arrows, have been employed to thin deer populations in many areas that would be otherwise inappropriate for hunters. Reports from Maryland alone show that local authorities, park rangers, USDA wildlife authorities, and private contractors have all utilized sharpshooters safely and with little disruption to residents or park goers. Sharpshooters were used to remove 12 does from a U.S. Navy facility in Calvert County, Maryland, as early as 1996. Corn is set out as bait to lure the deer to an area where sharpshooters await — ideally from an overhead ridge or treestand. This allows the shooter to selectively remove does (or impaired individuals) by shooting down at them. The ideal is for the deer to never know what hit them. Trapping and shooting deer has also been used, but this is much more stressful to deer, and makes it more difficult to thin the herd selectively. The cost of shooters is relatively high.

Another aspect of lethal deer control that has grown in recent years is that of donating the meat of killed deer to feed the hungry. Different groups have sprung up around the country, including the Venison Donation Coalition, in New York; Mississippi Sportsmen Against Hunger; Hunters Helping the Hungry in New Jersey; and Washington state's Sportsmen Against Hunger, to facilitate donations. Programs such as Hunters for the Hungry (whose Virginia branch donated 308,274 pounds of venison in 2003–2004) aid in coordinating the efforts of hunters and meat processors usually with little or no expense to the donating hunter for the processing. Hunters are instructed to field-dress carcasses and then deliver them to participating meat processors, leaving part or all of their kill for distribution to food banks, prisons, and others in need depending on the state and the individual program.

Immunocontraception

The newest approach to deer overpopulation involves slowing down the rate of reproduction. Nature does so when circumstances are particularly tough. Starvation and harsh winters often result in fewer fawns being born. Does may bear only one fawn instead of the usual twins, produce a stillborn, or in extreme circumstances abort or resorb fetuses. Wildlife biologists seek a way to spare the deer the stress of the natural method while achieving the same outcome, fewer fawns.

Birth control for wildlife began in the early 1970s in an effort to control the numbers of feral horses. Field testing in Nevada eventually resulted in success rates of more than 90 percent. In 1987, testing began on isolated deer populations.

One form of contraception, in which an immune response is used to prevent pregnancy, is called immunocontraception. So far the primary focus has been on a drug called Porcine Zona Pellucida (PZP), which contains a natural protein involved with the process of boar-sperm attachment to sow's eggs. The protein is recovered from pork-processing plants. When injected into a doe, this protein produces antibodies that attack the doe's own sperm-attachment proteins. Though the doe mates normally, an immune response from the doe's body prevents a buck's sperm from binding to the egg and conception does not occur. In addition to carrying no health risk for the deer, the proteins in the vaccine pose no threat to people, including those who eat venison from treated does. In one sense, it is undisputedly effective — anything injected with it, from mice to moose to man, will be sterilized. The challenge comes in getting the drug, a dose of which is only about 1 cc (about ¼ of a teaspoon), into the deer.

Several methods of administering the drug have been tested, from Pneu-Dart guns to blowguns, to something called a bio-bullet, which was designed to be "shot" into the deer with an air gun. This acts as a time-release capsule absorbed into a deer's system over several months. All of these have met with various degrees of success and frustration. Dart guns shoot out in a curved trajectory, which makes hitting what you aim for a challenge, especially in a wooded or brushy environment. In addition, up until recently, the darts were reportedly not consistent in the result, sometimes rendering a doe sterile for several years, other times having little effect. One product under testing, Spay-Vac, has a better track record with once-a-year administration. Blowguns are cheaper and easier to use, but have an even more unpredictable aim. The bio-bullet concept hasn't gotten off the ground because of the problem of injecting them with air guns. They are so lightweight that they don't travel well once airborne. Not enough air pressure, and the "bullets" bounce right off the doe's hide; too much can kill her. PZP is not simply fed to deer because there is no evidence to suggest it remains active when digested.

Contained populations offer the best opportunity to administer immunocontraception and to monitor the effects. Studies at Fire Island National Seashore in New York and the National

Institute of Standards and Technology (NIST) have been very enouraging. At Fire Island, where deer densities have been counted in excess of 200 per square mile, approximately 200 deer have been treated each year since 1998, leading to the discovery that two or more consecutive years of treatment cut fawning rates from between 85 and 90 percent. The population under study declined by 23 percent each year from 1998 through 2000. At the NIST, a peak population of 300 deer in 1997 droppped to 200 over the course of several years of treatment.

Another study on captive deer showed that, over the course of four years at the old Seneca Army Depot, the use of PZP produced an 85 percent decrease in fawning, according to a Cornell wildlife biologist. PZP is also being used on tule elk at Point Reyes National Seashore in California. In general, deer in these studies have been treated with two injections the first year, with a follow-up booster every year thereafter. Unfortunately, in large, free-roaming populations of deer, where deer are difficult to find and even more difficult to get close enough in order to dart, use of the vaccine may not be practical. Wildlife managers disagree, and testing continues.

Future Prospects

Many areas — particularly in northern New Jersey, parts of New York, much of New England, and in Missouri, Wisconsin, and California — suffer from severe deer pressure. In affected regions, the issue of deer control continues to be hotly debated, not only because of the damage deer inflict, but also because the deer are suffering — unthrifty in appearance, unhealthy, and starving. Indeed, in some areas the quality of the entire deer population is deteriorating. Average size is declining and bucks are producing smaller antlers. As ecologist and philosopher Aldo Leopold commented half a century ago, "There are no stags in the woods today like those whose antlers decorated the walls of feudal castles."

What has yet to be tried? Perhaps dominant bucks that have been rendered infertile could be introduced into areas of overpopulation. Where deer herds have deteriorated genetically due to inbreeding or environmental conditions, superior bucks could be relocated to strengthen faltering gene pools. Outlying areas, away from residential districts, could be opened up to deer dispersal by regulated burns, just as Native Americans did hundreds of years ago. The Humane Society of the United States (which maintains control of the use of PZP and has overseen the testing of it) even considered RU486 — the same "abortion pill" used by humans — for use in reducing deer pregnancies. The drug would be administered during the winter, when does are pregnant, and which also happens to be the easiest time to get near enough to deer to administer it.

"Soil, water, vegetation and wildlife cannot be managed separately for they are not separate entities but integral parts of the whole."

— Leonard Lee Rue III, *The Deer of North America*, 1989

Managing deer presents daunting challenges, not the least of which is balancing competing public interests and sentiments. The focus must be on the long-term welfare of the deer and of the communities in which they live. Healthy deer, in manageable numbers, can be a blessing and a benefit to all. And with population crises managed and adequate wild food available, we could concentrate on keeping a reasonable number of deer away from our landscape and garden plants.

Resources

Advocacy Groups

The Mule Deer Foundation
888-375-DEER
www.muledeer.org
Conservation of mule deer, black-tailed deer, and their habitats

Whitetails Unlimited
920-743-6777
www.whitetailsunlimited.com
Pro-hunting deer group

State Wildlife Agencies

Alabama Game and Fish
334-242-3465
www.dcnr.state.al.us/agfd

Alaska Dept. of Fish and Game
907-465-4100
www.adfg.state.ak.us

Arizona Game and Fish Dept.
602-942-3000
www.gf.state.az.us

Arkansas Game and Fish Dept.
800-364-4263
www.agfc.state.ar.us

California Dept. of Fish and Game
916-445-0411
www.dfg.ca.gov

Colorado Division of Wildlife
303-297-1192
http://wildlife.state.co.us

Connecticut Dept. of Environmental Protection
860-424-3000
http://dep.state.ct.us

Delaware Dept. of Natural Resources & Environmental Control
302-739-4403
www.dnrec.state.de.us

Florida Game and Fresh Water Fish
http://myfwc.com

Georgia Dept. of Natural Resources
770-414-3333
http://georgiawildlife.dnr.state.ga.us

Idaho Dept. of Fish and Game
208-334-3700
http://fishandgame.idaho.gov

Illinois Dept. of Conservation
217-782-6302
http://dnr.state.il.us

Indiana Dept. of Natural Resources
317-232-4080
www.in.gov/dnr/fishwild

Iowa Dept. of Natural Resources
515-281-5918
www.iowadnr.com

Kansas Dept. of Wildlife and Parks
620-672-5911
www.kdwp.state.ks.us

Kentucky Fish and Wildlife
800-858-1549
http://fw.ky.gov

Louisiana Dept. of Natural Resources
225-342-4540
http://dnr.louisiana.gov

Maine Dept. of Inland Fisheries and Wildlife
207-287-8000
www.state.me.us/ifw

Maryland Dept. of Natural Resources
410-260-8540
www.dnr.state.md.us/wildlife

Massachusetts Division of Fisheries and Wildlife
617-626-1590

Michigan Dept. of Natural Resources
517-373-1263
www.michigan.gov/dnr

Minnesota Dept. of Natural Resources
612-296-6157
www.dnr.state.mn.us

Mississippi Dept. of Wildlife Conservation
601-432-2400
www.mdwfp.com

Missouri Dept. of Conservation
573-751-4115
www.mdc.mo.gov

Montana Dept. of Fish, Wildlife, and Parks
406-444-2535
http://fwp.state.mt.us

Nebraska Game and Parks Commission
402-471-0641
www.ngpc.state.ne.us/wildlife

Nevada Dept. of Conservation and Natural Resources
775-687-4360
http://dcnr.nv.gov

New Hampshire Fish and Game Dept.
603-271-2461
www.wildlife.state.nh.us

New Jersey Dept. of Environment, Division of Fish and Wildlife
609-292-2965
www.state.nj.us/dep/fgw/

New Mexico Dept. of Game and Fish
505-476-8000
www.wildlife.state.nm.us

New York Bureau of Fish and Wildlife Services
518-402-8995
www.dec.state.ny.us/

North Carolina Wildlife Commission
919-733-7291
www.ncwildlife.org

North Dakota Game and Fish
701-328-6300
www.state.nd.us/gnf

Ohio Dept. of Natural Resources, Division of Wildlife
800-945-3543
www.dnr.state.oh.us/wildlife/

Oklahoma Dept. of Wildlife Conservation
405-521-3851
www.wildlifedepartment.com

Oregon Dept. of Fish and Wildlife
800-720-6339
www.dfw.state.or.us

Pennsylvania Game Commission
717-787-6286
www.pgc.state.pa.us

Rhode Island Dept. Environmental Management
401-222-6800
www.state.ri.us/dem/programs/bnatres/fishwild

South Carolina Dept. of Natural Resources
803-734-3886
www.dnr.state.sc.us

South Dakota Dept. of Game, Fish, and Parks
605-773-3381
www.sdgfp.info

Tennessee Wildlife Resources
615-781-6610
www.state.tn.us/twra

Texas Parks and Wildlife
512-389-4800
www.tpwd.state.tx.us

Utah Division of Wildlife Resources
801-538-4700
www.wildlife.utah.gov

Vermont Fish and Wildlife
802-241-3700
www.vtfishandwildlife.com

Virginia Dept. of Game and Inland Fisheries
804-367-1000
www.dgif.state.va.us

Washington Dept. of Wildlife
360-902-2200
http://wdfw.wa.gov

West Virginia Dept. of Wildlife Resources
304-558-2771
www.wvdnr.gov

Wisconsin Dept. of Natural Resources
608-266-6261
www.dnr.state.wi.us

Wyoming Dept. of Game and Fish
307-777-4600
http://gf.state.wy.us/

Internet Sites for Disease-Related Information

Centers for Disease Control and Prevention (CDC)
www.cdc.gov/ncidod/dvbid/lyme
Website for information on Lyme disease.

The American Lyme Disease Foundation
www.aldf.com

The Chronic Wasting Disease Alliance
www.cwd-info.org

Products and Suppliers

Type	Product	Active Ingredient(s)	Contact Information
Odor	Coast of Maine Salmon Plant Food	Processed fish parts	Coast of Maine Products www.coastofmaine.com 800-345-9315
Odor	DeerBusters Coyote Urine DeerBusters Professional Deer and Rabbit Repellent Deerbusters Professional Vegetable & Fruit Spray	Coyote Urine Ammonium salts of fatty acids Garlic	DeerBusters www.deerbusters.com 800-442-3337
Odor	Deer Chaser	Fatty acid soap & citrus	Gardeners Supply Company www.gardeners.com 888-833-1412
Odor	Deer No-No	85% soap, citrus	Deer No-No www.deernono.com 860-672-6264
Odor	Deer Out	Peppermint oil, white pepper, and garlic oil	Deer Out www.deerout.com 908-769-4242
Odor	Down to Earth Feather meal	Processed chicken feathers	Down to Earth Distributors www.down-to-earth.com/ 800-234-5932
Odor	ferti-lome Rabbit and Deer Repellent	Ammonium soaps of higher fatty acids	Voluntary Purchasing Groups www.v-p-g.com
Odor	Hinder	Ammonium soaps of higher fatty acids	Peaceful Valley Farm Supply www.groworganic.com 888-784-1722
Odor	Liquid Fence	Putrescent egg and garlic	Liquid Fence www.liquidfence.com 888-923-3623
Odor	Master Gardeners Deer and Insect Repellent	Garlic	Master Gardening Products www.mastergardening.com 888-422-3337
Odor	Milorganite	Treated sewage	Milorganite www.milorganite.com 800-304-6204
Odor	Organic Gem Fish Fertilizer	Processed fish parts	Advanced Marine Technologies www.organicgem.org 508-991-5225
Odor	Plantskydd	Dried bloodmeal & processed animal protein	Tree World www.plantskydd.com 800-252-6051
Odor	Shake Away Deer and Large Animal Repellent	Coyote Urine	Shake Away www.shake-away.com 800-517-9207

Type	Active Ingredient(s)	Product	Contact Information
Odor/ Taste	Not Tonight Deer!	Dehydrated whole egg solids; white pepper	Not Tonight Deer! www.nottonight.com
Odor/ Taste	Deer Away (Big Game Repellent)	Putrescent egg solids; extracts of chili and mustard oil	www.treehelp.com 877-356-7333
Odor/ Taste	Deerbusters I (liquid or powder)	Putrescent egg, garlic, and capsaicin	www.deerbusters.com 800-248-3337
Odor/ Taste	Deer-Off	Putrescent egg and capsicum	www.deer-off.com 800-DEER-OFF
Odor/ Taste	Bonide Shot-Gun Deer and Rabbit Repellent	Egg, garlic, and pepper	Bonide Products www.bonideproducts.com 315-736-8231
Odor/ Taste	Bobbex Deer Repellent	Garlic oil, acetic acid, cloves, fish meal, edible fish oil, onions, eggs, vanillin, wintergreen oil	Bobbex www.bobbex.com 800-792-4449
Odor/ Taste	Repellex	Dried blood, paprika	Repellex www.repellex.com 877-REPEL-IT
Taste	Millers Hot Sauce Game Repellent	2.5% capsaicin	Miller Chemical www.millerchemical.com
Taste	Scoot Deer	Castor oil and capsaicin	Bird-X Inc. www.bird-x.com 312-226-2473
Taste	N.I.M.B.Y.	Castor oil and capsaicin	DMX Industries www.dmxind.com 314-385-0076
Taste	Hot Pepper Wax Animal Repellent	Cayenne peppers, assorted repelling herbs, and food-grade paraffin wax	Hot Pepper Wax Inc. www.hotpepperwax.com, 800-627-6840
Taste	Nott's Chew-Not	Thiram	Nott Manufacturing 914-635-3243
Systemic	Ro-Pel Systemic Tablets Repellex Systemic ML2	Bitrex 12% Denatonium Benzoate	Burlington Scientific Corp. www.ropel.com 631-694 4700
Sound/ Touch	Scarecrow Deer and Animal Repeller	Sprinkler	www.scatmat.com 800-767-8658
Physical barrier	Deer-X Netting	Deer netting	Dalen Products www.gardeneer.com 865-966-3256
Physical barrier	Benner's Deer Fencing	Deer netting	Benner's Gardens www.bennersgardens.com 800-BIG-DEER

INDEX

Page numbers in **bold** indicate tables; page numbers in *italics* indicate illustrations.

Other Storey Titles You Will Enjoy

The Fence Bible, by Jeff Beneke.
A complete resource to build fences that enhance the landscape while fulfilling basic functions.
272 pages. Paper. ISBN-13: 978-1-58017-530-2.
Hardcover with jacket. ISBN-13: 978-1-58017-586-9.

Fences for Pasture & Garden, by Gail Damerow.
Sound, up-to-date advice and instruction to make building fences a task anyone can tackle with confidence.
160 pages. Paper. ISBN-13: 978-0-88266-753-9.

The Gardener's Bug Book, by Barbara Pleasant.
The health-conscious gardener's guide to safely reduce pests while producing bountiful, environmentally safe, and chemical-free harvests.
144 pages. Paper. ISBN-13: 978-0-88266-609-9.

The Gardener's Weed Book, by Barbara Pleasant.
Complete coverage of the pros and cons of weeds, proven methods for controlling unwanted ones, and illustrations of common varieties.
208 pages. Paper. ISBN-13: 978-0-88266-921-2.

Incredible Vegetables from Self-Watering Containers, by Edward C. Smith.
A foolproof method to produce a bountiful harvest without the trouble of a traditional earth garden.
256 pages. Paper. ISBN-13: 978-1-58017-556-2.
Hardcover. ISBN-13: 978-1-58017-557-9.

The Vegetable Gardener's Bible, by Edward C. Smith.
A reinvention of vegetable gardening that shows how to have your most successful garden ever.
320 pages. Paper. ISBN-13: 978-1-58017-212-7.